T0332714

Springer Tracts in Natural Philosophy

Volume 32

Volume 32

Edited by C. Truesdell

Springer Tracts in Natural Philosophy

William Alan Day

A Commentary on Thermodynamics

Springer Science+Business Media, LLC

William Alan Day
Mathematical Institute
University of Oxford
24–29 St Giles
Oxford OX1 3LB
England

AMS Classification: 80AB (80A10)

Library of Congress Cataloging-in-Publication Data
Day, William Alan.
    A commentary on thermodynamics.
    (Springer tracts in natural philosophy;v.32)
    Bibliography: p.
    Includes index.
    1. Thermodynamics. 2. Thermoelasticity. I. Title.
II. Series.
QC311.D36   1988   536'.7   87-23360

Typeset by Asco Trade Typesetting Ltd., Hong Kong.

9 8 7 6 5 4 3 2 1

ISBN 978-0-387-96615-1   ISBN 978-1-4419-8550-7 (eBook)
DOI 10.1007/978-1-4419-8550-7

Contents

Introduction

The purpose of this tract is to assemble a number of comments on thermodynamics.

Mathematicians who take a professional interest in thermodynamics usually find themselves exasperated at the obscurity and imprecision which bedevil traditional ways of presenting that subject. One way of attempting to remedy matters, and one which has an obvious attraction, is to follow the pattern of Euclid's treatment of geometry and present thermodynamics as a formal deductive system of lemmas, propositions, and theorems proceeding from clearly stated axioms.

It is not my intention to pursue such a course, for I believe that much obscurity can be removed in an alternative way, namely by setting thermodynamics in the context of field theories which incorporate the effects of heat conduction and inertia, and proving appropriate results about the partial differential equations which arise in those theories. Indeed, I would go further and say that it is only the latter procedure which can yield real insight into, let us say, what it means for a process to be "quasi-static", "reversible", or "irreversible". I would add too that the latter procedure is likely to turn out to be both more interesting and more difficult, in terms of mathematical content, than is the construction of an axiomatic scheme.

In accordance with these beliefs, I propose to comment on thermodynamics from the standpoint of four theories of thermoelasticity;

one of the theories is exact but the other three are approximate at best.

Questions of geometry, frame-indifference, and material symmetry, which become important in higher dimensions, are excluded by the device of treating only the simplest of cases, namely that of a one-dimensional body.

I shall begin by deriving the nonlinear theory of thermoelasticity, which is an exact theory. It will be found that the displacement $u(x, t)$ and the temperature $T(x, t)$, which depend upon a single spatial variable x and the time t only, must satisfy partial differential equations of the form

$$\frac{\partial}{\partial x}\left[\frac{\partial \hat{F}}{\partial E}\left(\frac{\partial u}{\partial x}, T, x\right)\right] + f = \rho\frac{\partial^2 u}{\partial t^2},$$

$$\frac{\partial}{\partial x}\left[\hat{q}\left(\frac{\partial u}{\partial x}, T, \frac{\partial T}{\partial x}, x\right)\right] + h = -T\frac{\partial}{\partial t}\left[\frac{\partial \hat{F}}{\partial T}\left(\frac{\partial u}{\partial x}, T, x\right)\right].$$

Although, as will be seen, some progress can be made with these equations, it is not possible to give a comprehensive account of them at the present time. That being so, it is natural to introduce various simplifications or approximations: one such set of simplifications and approximations leads to homogeneous and dissipationless thermoelasticity, while another set leads to linearized thermoelasticity. In the closing stages of the tract it will be necessary to examine a fourth theory—a slight generalization of linearized thermoelasticity which I shall call nonstandard linearized thermoelasticity.

Homogeneous and dissipationless thermoelasticity retains some of the nonlinear features of the exact theory. On the other hand, it ignores heat conduction and inertia, and so drastic are the attendant approximations that it is a field theory in only a rather trivial sense.

Linearized thermoelasticity, by contrast, is a genuine field theory, and one which takes account of heat conduction and inertia. However, the nonlinearity which is integral to the exact theory is sacrificed in favour of having to deal with linear differential equations.

It is inevitable that each of the approximate theories should distort some aspects of the truth, and examples of such distortions will be encountered later. The approximate theories deserve to be studied nevertheless, for we cannot understand the exact theory without first having understood the approximate theories to which it gives rise.

Notation is a troublesome matter. The modern school of rational thermodynamics adopts a system which agrees, in the main, with that of Gibbs and differs from what is commonly found in textbooks on classical thermodynamics. In the hope of making the tract accessible to a wider readership I use, where possible, notation which is close to that of the textbooks. It is impossible to be entirely consistent in such matters, though, and there are instances in which it is essential not to conform to standard presentations. Thus, differentials find no place here, and statements that are commonly framed in terms of differentials are replaced by statements about the derivatives to which they really refer. I may add that the theory of differential forms, which some analysts regard as the language in which any discussion of thermodynamics ought to be conducted, is of little assistance in the context of this tract.

CHAPTER 1

Nonlinear Thermoelasticity

1. I propose to begin by deriving the exact nonlinear theory of one-dimensional thermoelasticity.

My method is, in part at least, bald assertion of the equations that are presumed to be in force. Thus, I make no attempt to deduce the equations of state from more primitive assumptions; that way of proceeding would involve commitment to some specific statement of a second law of thermodynamics—the Clausius–Duhem inequality being an example. For the limited purposes of this tract, such commitment is unnecessary.

2. It may be helpful to think of a one-dimensional and deformable body as being an idealization of a three-dimensional slab which undergoes a motion in which particles of the slab are displaced in a direction orthogonal to the faces of the slab. Thus, let $Oxyz$ be a system of rectangular Cartesian coordinates and suppose that, when it is placed in one of its configurations (the reference configuration), the slab occupies a region

$$\{(x, y, z): A \leqq x \leqq B, -\infty < y < \infty, -\infty < z < \infty\}$$

and is bounded by the planes $x = A$ and $x = B$.

If $u(x, t)$ is the displacement field, the particle which is at the point

$$(x, y, z),$$

when the slab is in the reference configuration, comes to occupy the point

$$(x + u(x, t), y, z)$$

at the time t.

It should be noted that it is not necessarily the case that

$$A \leqq x + u(x, t) \leqq B.$$

Moreover, the thickness of the slab, which is

$$B - A + u(B, t) - u(A, t),$$

will vary with t.

Such vector fields as the external force and the heat flux are all, like the displacement, presumed to be aligned along the x-axis.

3. These introductory sentences apart, the letters x, y, z will not, in fact, be used to denote Cartesian coordinates, which now disappear from view. Henceforth x, y, z are points of the closed and bounded interval

$$[A, B].$$

In the language of continuum mechanics, these points are material coordinates, or Lagrangian coordinates.

It is convenient to refer to the interval $[A, B]$ as the body, and to the points A and B as the boundary points. By a subbody is meant any closed and bounded subinterval of $[A, B]$.

In what follows all fields depend, at most, upon the point x and the time t, which, it is understood, lies in a closed and bounded interval

$$[t_1, t_2].$$

Thus, fields such as

$$\xi(x), \quad \phi(t), \quad \psi(x, t)$$

are real-valued, or sometimes complex-valued, functions that are defined, respectively, on the sets

$$[A, B], \quad [t_1, t_2], \quad [A, B] \times [t_1, t_2].$$

If the domain of definition is clear, x or t may be omitted, and the fields referred to as plain

$$\xi, \quad \phi, \quad \psi.$$

The net increase in $\xi(x)$ on a subbody $[a, b]$ is

$$[\xi]_a^b = \xi(b) - \xi(a)$$

and, likewise, the net increase in $\phi(t)$ on the interval $[t_1, t_2]$ is

$$[\phi]_{t_1}^{t_2} = \phi(t_2) - \phi(t_1).$$

Net increases in real-valued fields may well be negative, of course.

Spatial derivatives are denoted by dashes, and temporal derivatives are denoted by dots; thus

$$\xi'(x) = \frac{d}{dx}\xi(x), \qquad \psi'(x, t) = \frac{\partial}{\partial x}\psi(x, t),$$

$$\dot{\phi}(t) = \frac{d}{dt}\phi(t), \qquad \dot{\psi}(x, t) = \frac{\partial}{\partial t}\psi(x, t).$$

The usual descriptions

$$C^0, \quad C^1, \quad C^2, \quad C^\infty$$

are reserved for fields that are, respectively, continuous, continuously differentiable, twice continuously differentiable, and continuously differentiable infinitely often.

4. Kinematics is readily disposed of in a one-dimensional context. The displacement

$$u(x, t)$$

is required to be C^2 and to preserve the strict ordering of points, that is to say

$$x + u(x, t) < y + u(y, t)$$

whenever x and y belong to $[A, B]$ and $x < y$.

The latter requirement will be met by supposing slightly more, namely that the spatial derivative of the field

$$x + u(x, t)$$

is positive. In terms of the strain

$$E = u'$$

this is tantamount to supposing that

$$1 + E > 0.$$

Other derivatives of the displacement that play important roles are

$$\dot{u}, \quad \ddot{u}, \quad \dot{u}' \quad (=\dot{E})$$

which are, respectively, the velocity, the acceleration, and the strain-rate.

5. In order to formulate balance laws for momentum and energy it is necessary to introduce:

the mass density in the reference configuration $\rho(x)$,
the stress $\sigma(x, t)$,
the external force density $f(x, t)$,
the internal energy density $U(x, t)$,
the heat flux $q(x, t)$,
the external rate-of-heating density $h(x, t)$.

Of these ρ, f, h are required to be C^0, while σ, U, q are required to be C^1.

Each of the densities ρ, U, f, h is measured with respect to length in the reference configuration. Thus, for any subbody $[a, b]$,

$\int_a^b \rho \, dx = $ mass of $[a, b]$,
$\int_a^b U \, dx = $ internal energy of $[a, b]$,
$\int_a^b f \, dx = $ force exerted on $[a, b]$ by the exterior of the body,
$\int_a^b h \, dx = $ rate of heating of $[a, b]$ by the exterior of the body.

It may be remarked, although this fact is not needed subsequently, that the densities of mass, internal energy, external force, and external rate of heating, measured with respect to length in the deformed configuration, can be obtained from

$$\rho, \quad U, \quad f, \quad h,$$

respectively, by dividing them by the positive factor $1 + E$.

The following terms figure in the balance laws:

$[\sigma]_a^b = $ net contact force on $[a, b]$,
$[\sigma\dot{u}]_a^b = $ net rate of working of the contact forces on $[a, b]$,
$[q]_a^b = $ net heat flux into $[a, b]$,
$\int_a^b \rho\dot{u} \, dx = $ momentum of $[a, b]$,
$\frac{1}{2}\int_a^b \rho\dot{u}^2 \, dx = $ kinetic energy of $[a, b]$,
$\int_a^b f\dot{u} \, dx = $ rate of working of the external force on $[a, b]$.

The convention adopted here, which makes $[q]_a^b$ the net heat flux *into* $[a, b]$, means that q is the negative of what is normally called the heat flux. This departure from normal practice is justified on the ground that it removes a number of negative signs at later stages.

It is now possible to postulate that:

For each subbody $[a, b]$, the momentum balance law

$$[\sigma]_a^b + \int_a^b f\, dx = \overline{\int_a^b \rho\dot{u}\, dx}$$

is satisfied, and so is the energy balance law

$$[\sigma\dot{u}]_a^b + \int_a^b f\dot{u}\, dx + [q]_a^b + \int_a^b h\, dx = \tfrac{1}{2}\overline{\int_a^b \rho\dot{u}^2\, dx} + \int_a^b U\, dx.$$

6. The first conclusion to be drawn is:

The balance laws imply the momentum equation

$$\sigma' + f = \rho\ddot{u},$$

the energy equation

$$(\sigma\dot{u})' + f\dot{u} + q' + h = \rho\dot{u}\ddot{u} + \dot{U},$$

and the first reduced energy equation

$$\sigma\dot{E} + q' + h = \dot{U}.$$

The verification is straightforward. For,

$$[\sigma]_a^b = \int_a^b \sigma'\, dx,$$

$$\overline{\int_a^b \rho\dot{u}\, dx} = \int_a^b \rho\ddot{u}\, dx,$$

and, therefore, the momentum balance law implies that

$$\int_a^b (\sigma' + f - \rho\ddot{u})\, dx = 0$$

for every subbody $[a, b]$. Since the integrand is C^0 the momentum equation must hold.

What is essentially the same argument shows that the energy balance law implies the energy equation.

Finally, the momentum equation, the energy equation, and the identity

$$(\sigma \dot{u})' = \sigma' \dot{u} + \sigma \dot{E}$$

imply the first reduced energy equation.

7. Granted the smoothness of the fields involved, the equations just derived exhaust the content of the balance laws.

Let $u(x, t)$ be C^2, let $\sigma(x, t)$ and $q(x, t)$ be C^1, let $\rho(x)$, $f(x, t)$, $h(x, t)$ be C^0, and let $E = u'$. Furthermore, let the momentum equation and the energy equation or, equivalently, the momentum equation and the first reduced energy equation, be satisfied. Then the momentum balance law and the energy balance law are valid for each subbody $[a, b]$.

8. The next task is to relate the stress, the internal energy density, and the heat flux, to the displacement and the temperature.

No attempt will be made to derive the relations from some version of a second law, such as the Clausius–Duhem inequality.† The considerations of this tract are independent of whether or not that inequality is a correct statement of the second law. It should be noted, though, that, as will be established in §45, the Clausius–Duhem inequality is certainly valid within the theory of nonlinear thermoelasticity constructed in this chapter.

I introduce, without explanation:
the absolute temperature $T(x, t)$,
the temperature gradient $g = T'$,
the entropy density $S(x, t)$,
the free energy density $F(x, t)$.
The first of these is required to be positive, thus

$$T > 0.$$

---

† Coleman and Noll [3] were the first to adopt the Clausius–Duhem inequality as the foundation stone for a rational and systematic theory of continuum thermodynamics. Other accounts of modern continuum thermodynamics are to be found in Coleman and Owen [4], Day [5], Müller [8], Owen [9], Serrin [12], Truesdell [14], and Truesdell and Bharatha [15]. In some of those accounts the Clausius–Duhem inequality does not play as fundamental a role as it does in the theory of Coleman and Noll.

The adjective "absolute" will be omitted henceforth, T being described simply as the temperature.

The constitutive relations embody the idea that the stress, the internal energy density, the entropy density, and the free energy density, as a point x and a time t, are determined by the strain and the temperature at that point and that time; the heat flux, however, depends upon the temperature gradient as well.

In formal terms, there exist:

the stress response function $\hat{\sigma}(E, T, x)$,
the internal energy response function $\hat{U}(E, T, x)$,
the entropy response function $\hat{S}(E, T, x)$,
the free energy response function $\hat{F}(E, T, x)$,
the heat flux response function $\hat{q}(E, T, g, x)$,

and these determine the stress, the internal energy density, the entropy density, the free energy density, and the heat flux, through the constitutive relations

$$\sigma(x, t) = \hat{\sigma}(E(x, t), T(x, t), x),$$

$$U(x, t) = \hat{U}(E(x, t), T(x, t), x),$$

$$S(x, t) = \hat{S}(E(x, t), T(x, t), x),$$

$$F(x, t) = \hat{F}(E(x, t), T(x, t), x),$$

$$q(x, t) = \hat{q}(E(x, t), T(x, t), g(x, t), x).$$

The explicit dependence of each of the response functions upon the material coordinate x means that the response is permitted to be inhomogeneous.

9. The first four constitutive relations are far from being independent of each other and, indeed, it is postulated that:

The free energy response function is C^2 in its dependence upon E and T, and it determines the response functions for the stress, the internal energy density, and the entropy density through the equations of state

$$\hat{\sigma} = \partial_E \hat{F},$$

$$\hat{U} = \hat{F} - T\partial_T \hat{F},$$

$$\hat{S} = -\partial_T \hat{F}.$$

The equations of state imply that:

Maxwell's relation

$$\partial_T \hat{\sigma} + \partial_E \hat{S} = 0$$

holds, and so do the equation

$$U = F + TS$$

and Gibbs's relation

$$\dot{U} = \sigma \dot{E} + T \dot{S}.$$

The first two assertions are immediate, while Gibbs's relation is a consequence of the equation

$$\dot{U} = \dot{F} + \dot{T} S + T \dot{S}$$

and the observation that

$$\dot{F} = (\partial_E \hat{F}) \dot{E} + (\partial_T \hat{F}) \dot{T} = \sigma \dot{E} - S \dot{T}.$$

It is now clear that:

Gibbs's relation and the first reduced energy equation imply the second reduced energy equation

$$q' + h = T \dot{S},$$

which, since T is positive, is equivalent to

$$\dot{S} = \left(\frac{q}{T}\right)' + \frac{gq}{T^2} + \frac{h}{T}.$$

10. It will be essential to restrict the heat flux by means of the postulate:

The heat flux response function is C^2 in its dependence upon g, and is positive or negative according as g is positive or negative, that is to say the heat conduction inequality

$$g \hat{q}(E, T, g, x) > 0$$

holds whenever $g \neq 0$.

Two important consequences will be noted at this stage; the second concerns the thermal conductivity response function, which is

$$\hat{k}(E, T, x) = \partial_g \hat{q}(E, T, 0, x).$$

The heat conduction inequality implies that $q = 0$ whenever $g = 0$, that is to say

$$\hat{q}(E, T, 0, x) = 0,$$

and that

$$\hat{k}(E, T, x) \geqq 0.$$

In order to arrive at these conclusions it is enough to note that, for fixed values of E, T, x, the function

$$g \to g\hat{q}(E, T, g, x)$$

attains its minimum value at $g = 0$. Thus, the first derivative

$$\partial_g(g\hat{q}) = \hat{q} + g\,\partial_g\hat{q}$$

must vanish at $g = 0$, and the second derivative

$$\partial_g^2(g\hat{q}) = 2\,\partial_g\hat{q} + g\,\partial_g^2\hat{q}$$

cannot be negative at $g = 0$. The desired conclusions follow immediately.

In fact, more is true than has been claimed. For, when the temperature gradient is small, the behaviour of the heat flux is described by the order relation

$$\hat{q}(E, T, g, x) = \hat{k}(E, T, x)g + o(|g|) \qquad \text{when} \quad g \to 0.$$

Neglect of the term $o(|g|)$ would produce Fourier's law of heat conduction

$$q = kg,$$

with a thermal conductivity k which is nonnegative and may depend upon strain, temperature, and position in the body. (In view of the way q has been defined, there is no minus sign on the right-hand side of Fourier's law.)

11. The partial differential equations advertised in the Introduction follow from the hypotheses introduced thus far. For, on substituting from the equations of state for the stress and the entropy, and from the constitutive relation for the heat flux, into the momentum equation and the second reduced energy equation, and remembering that $E = u'$ and $g = T'$, we obtain the conclusion:

Within nonlinear thermoelasticity, the displacement and temperature fields satisfy the nonlinear displacement–temperature equations

$$[\partial_E \hat{F}(u', T, x)]' + f = \rho \ddot{u},$$
$$[\hat{q}(u', T, T', x)]' + h = -T\overline{[\partial_T \hat{F}(u', T, x)]}\,\dot{}\,.$$

Before attempting to draw any conclusions from the nonlinear theory, I turn to two simplified and approximate theories to which it gives rise.

The first of these, namely homogeneous and dissipationless thermoelasticity, is essentially classical thermodynamics restricted to reversible processes. The reader is urged to consult Truesdell's history of thermodynamics in the formative period 1822–1854 [13].

The second simplified and approximate theory is linearized thermoelasticity, a fuller account of which is to be found in Carlson's article [1].

CHAPTER 2

Simplification and Approximation

Part 1. Homogeneous and Dissipationless Thermoelasticity

12. Homogeneous and dissipationless thermoelasticity rests upon four simplifying or approximating assumptions.

The first simplification is:

The material response is homogeneous.

In more detail, the mass density

$$\rho(x)$$

is constant, and the response functions

$$\hat{F}(E, T, x), \quad \hat{q}(E, T, g, x)$$

for the free energy and the heat flux are independent of x.

In view of the equations of state, the response functions

$$\hat{\sigma}(E, T, x), \quad \hat{U}(E, T, x), \quad \hat{S}(E, T, x)$$

for the stress, the internal energy, and the entropy must also be independent of x.

The second simplification is:

The strain, temperature, and external rate-of-heating density are spatially homogeneous.

This means that

$$E(x, t), \quad T(x, t), \quad h(x, t)$$

are independent of x and may be written more succinctly as

$$E(t), \quad T(t), \quad h(t).$$

Taken together, the first two simplifying assumptions force

$$F(x, t), \quad \sigma(x, t), \quad U(x, t), \quad S(x, t)$$

to be independent of x. They can, accordingly, be written as

$$F(t), \quad \sigma(t), \quad U(t), \quad S(t).$$

The disappearance of all dependence upon x means that homogeneous and dissipationless thermoelasticity is a field theory in only a rather trivial sense.

The second simplification has the further, and very important, consequence that the temperature gradient $g = 0$. In the light of §10, it must be that the heat flux

$$q = 0.$$

In brief: heat conduction is absent from homogeneous and dissipationless thermoelasticity.

The heat conduction inequality, it should be noted, is true vacuously, since within this theory it is never the case that $g \neq 0$.

Since $q = 0$, the second reduced energy equation collapses to

$$h = T\dot{S},$$

and provides the means of calculating the external rate-of-heating density which the exterior of the body must supply in order to maintain the homogeneous strain and temperature fields.

The third simplification is:

The external force density $f = 0$.

The fourth assumption is an approximation rather than a simplification:

The inertial term

$$\rho\ddot{u}$$

may be omitted from the momentum equation.

It is this last approximation which underlies the belief that classical thermodynamics deals only with "quasi-static" processes.

Because of the third and fourth assumptions, the momentum equation reduces to the equation

$$\sigma' = 0,$$

and this is satisfied automatically since the stress is spatially homogeneous.

As in the nonlinear theory, it is required that:

The free energy response function $\hat{F}(E, T)$ is C^2 and it determines the response functions for the stress, the internal energy density, and the entropy density, through the equations of state

$$\hat{\sigma} = \partial_E \hat{F},$$

$$\hat{U} = \hat{F} - T \, \partial_T \hat{F},$$

$$\hat{S} = -\partial_T \hat{F}.$$

The following conclusions are not upset by our assumptions:

Maxwell's relation

$$\partial_T \hat{\sigma} + \partial_E \hat{S} = 0,$$

the equation

$$U = F + TS,$$

and Gibbs's relation

$$\dot{U} = \sigma \dot{E} + T \dot{S}$$

all remain valid within homogeneous and dissipationless thermoelasticity.

Part 2. Linearized Thermoelasticity

13. Linearized thermoelasticity is based upon a different set of simplications and approximations from those of homogeneous and dissipationless thermoelasticity. The material response need not be homogeneous in linearized thermoelasticity, and the strain and temperature fields need not be spatially homogeneous. Thus, we are confronted by a genuine, if approximate, field theory. Moreover, the temperature gradient and the heat flux need not vanish, and heat conduction within the body itself, and between the body and its exterior, are now important effects. Inertia too is important because the term

$$\rho \ddot{u}$$

is retained in the momentum equation.

The first simplification is:

There is a positive constant T_0 such that $\hat{\sigma}(0, T_0, x)$ is independent of x.

Henceforth, T_0 is described as the reference temperature, and the constant

$$\sigma_0 = \hat{\sigma}(0, T_0, x)$$

is described as the residual stress; σ_0 is the stress which results when the body is maintained in the reference configuration, and at the reference temperature, that is when $u = 0$ and $T = T_0$.

Since, as §10 tells us,

$$\hat{q}(0, T_0, 0, x) = 0,$$

the nonlinear displacement–temperature equations (§11) admit the static solution

$$u(x, t) = 0, \qquad T(x, t) = T_0,$$

provided that the external force density f and the external rate-of-heating density h both vanish.

14. The two approximations embody what is the characteristic assumption of the linearized theory, namely that u and T, and their spatial derivatives, remain close to the values they take in the static solution. Thus, the strain $E\ (=u')$, the temperature difference $T - T_0$, and the temperature gradient $g\ (=T')$ are all small.

The first approximation is:

In the second reduced energy equation, the term

$$T\dot{S}$$

may be replaced by

$$T_0\dot{S},$$

and, therefore, the approximate second reduced energy equation

$$q' + h = T_0 \dot{S}$$

is presumed to be in force.

It will be found that this approximation, which is dictated by considerations of mathematical convenience, leads to some distortion of the true state of affairs as described by the nonlinear theory.

15. If $E, T - T_0, g$ do remain small, it is natural to attempt to approximate the equations of state for the stress and the entropy density, and the constitutive relation for the heat flux, by way of the relations

$$\hat{\sigma}(E, T, x) = \sigma_0 + \partial_E \hat{\sigma}(0, T_0, x)E + \partial_T \hat{\sigma}(0, T_0, x)(T - T_0),$$

$$\hat{S}(E, T, x) = \hat{S}(0, T_0, x) + \partial_E \hat{S}(0, T_0, x)E + \partial_T \hat{S}(0, T_0, x)(T - T_0),$$

$$\hat{q}(E, T, g, x) = \hat{k}(0, T_0, x)g.$$

By virtue of Maxwell's relation (§9), two of the coefficients on the right-hand sides are connected by the equation

$$\partial_T \hat{\sigma}(0, T_0, x) + \partial_E \hat{S}(0, T_0, x) = 0.$$

The following notations and names are employed for the coefficients:

the residual entropy density $S_0(x)$ $= \hat{S}(0, T_0, x)$,
the isothermal elastic modulus $\beta(x)$ $= \partial_E \hat{\sigma}(0, T_0, x)$,
the stress–temperature modulus $\mu(x)$ $= -\partial_T \hat{\sigma}(0, T_0, x)$,
$= \partial_E \hat{S}(0, T_0, x)$,
the specific heat at constant strain $c(x) = T_0 \partial_T \hat{S}(0, T_0, x)$,
the thermal conductivity $k(x)$ $= \hat{k}(0, T_0, x)$.

In this way, we arrive at the second, and final, approximation:

The equations of state for the stress and the entropy density, and the constitutive relation for the heat flux, may be replaced by:

$$\hat{\sigma}(E, T, x) = \sigma_0 + \beta(x)E - \mu(x)(T - T_0),$$

$$\hat{S}(E, T, x) = S_0(x) + \mu(x)E + \frac{c(x)}{T_0}(T - T_0),$$

$$\hat{q}(E, T, g, x) = k(x)g.$$

In the present context:

The heat conduction inequality

$$gq > 0 \qquad if \quad g \neq 0$$

is satisfied if and only if

$$k(x) > 0 \quad in \ [A, B],$$

and this last restriction is presumed to be in force.

16. If we substitute from the approximate equations of state, and the approximate constitutive relation for the heat flux, into the momentum equation and the approximate second reduced energy equation, and remember that $E = u'$ and $g = T'$, we obtain the conclusion:

Within linearized thermoelasticity the displacement and temperature fields satisfy the linearized displacement–temperature equations

$$[\beta u' - \mu(T - T_0)]' + f = \rho \ddot{u},$$

$$[kT']' + h = T_0 \mu \dot{u}' + c\dot{T}.$$

The inequalities

$$1 + E > 0, \qquad T > 0,$$

of the nonlinear theory play no part in the linearized theory; they are satisfied, of course, if, as has been supposed, E is sufficiently small and T is sufficiently close to the reference temperature T_0.

17. In the context of the linearized theory it is convenient to *define* the free energy response function

$$\hat{F}(E, T, x)$$

so as to satisfy the relations

$$\partial_E \hat{F} = \hat{\sigma} = \sigma_0 + \beta E - \mu(T - T_0),$$

$$-\partial_T \hat{F} = \hat{S} = S_0 + \mu E + \frac{c}{T_0}(T - T_0).$$

That definition forces $\hat{F}$ to have the form

$$\hat{F}(E, T, x) = F_0(x) + \sigma_0 E - S_0(x)(T - T_0)$$

$$+ \tfrac{1}{2}\beta(x)E^2 - \mu(x)E(T - T_0) - \frac{c(x)}{2T_0}(T - T_0)^2,$$

where

$$F_0(x) = \hat{F}(0, T_0, x)$$

is the residual free energy density.

It is also convenient to define the internal energy response function

$$\hat{U}(E, T, x)$$

according to the rule

$$\hat{U} = \hat{F} + T\hat{S}.$$

Thus

$$\hat{U}(E, T, x) = U_0(x) + (\sigma_0 + \mu(x)T_0)E + c(x)(T - T_0)$$
$$+ \tfrac{1}{2}\beta(x)E^2 + \frac{c(x)}{2T_0}(T - T_0)^2,$$

where

$$U_0(x) = \hat{U}(0, T_0, x) = F_0(x) + T_0 S_0(x)$$

is the residual internal energy density.

The purpose of these definitions is to ensure that:

Gibbs's relation

$$\dot{U} = \sigma\dot{E} + T\dot{S}$$

remains valid within linearized thermoelasticity.

Indeed, direct calculation yields

$$\dot{U} = (\sigma_0 + \mu T_0)\dot{E} + c\dot{T} + \beta E\dot{E} + \frac{c}{T_0}(T - T_0)\dot{T}$$

$$= (\sigma_0 + \mu T_0 + \beta E)\dot{E} + \frac{c}{T_0}T\dot{T},$$

and

$$\sigma\dot{E} + T\dot{S} = (\sigma_0 + \beta E - \mu(T - T_0))\dot{E} + T\left(\mu\dot{E} + \frac{c}{T_0}\dot{T}\right)$$

$$= (\sigma_0 + \mu T_0 + \beta E)\dot{E} + \frac{c}{T_0}T\dot{T},$$

and, hence, the assertion is correct.

Synopsis

18. At this point I summarize, for the sake of comparison, the three theories with which the major part of the tract is concerned.

Nonlinear Thermoelasticity

$$E = u', \qquad g = T',$$

$$\sigma' + f = \rho \ddot{u},$$

$$q' + h = T\dot{S},$$

$$\sigma = \partial_E \hat{F}(E, T, x),$$

$$S = -\partial_T \hat{F}(E, T, x),$$

$$U = F + TS,$$

$$q = \hat{q}(E, T, g, x),$$

$$gq > 0 \qquad \text{if} \quad g \neq 0.$$

Homogeneous and Dissipationless Thermoelasticity

$$E = E(t), \qquad T = T(t),$$

$$h = T\dot{S},$$

$$\sigma = \partial_E \hat{F}(E, T),$$

$$S = -\partial_T \hat{F}(E, T),$$

$$U = F + TS,$$

$$f = g = q = 0.$$

Linearized Thermoelasticity

$$E = u', \qquad g = T',$$

$$\sigma' + f = \rho \ddot{u},$$

$$q' + h = T_0 \dot{S},$$

$$\hat{F}(E, T, x) = F_0(x) + \sigma_0 E - S_0(x)(T - T_0)$$

$$+ \tfrac{1}{2}\beta(x)E^2 - \mu(x)E(T - T_0) - \frac{c(x)}{2T_0}(T - T_0)^2,$$

$$\sigma = \partial_E \hat{F} = \sigma_0 + \beta(x)E - \mu(x)(T - T_0),$$

$$S = -\partial_T \hat{F} = S_0(x) + \mu(x)E + \frac{c(x)}{T_0}(T - T_0),$$

$$\hat{U}(E, T, x) = \hat{F} + T\hat{S}$$

$$= U_0(x) + (\sigma_0 + \mu(x)T_0)E + c(x)(T - T_0)$$

$$+ \tfrac{1}{2}\beta(x)E^2 + \frac{c(x)}{2T_0}(T - T_0)^2,$$

$$q = k(x)g,$$

$$k(x) > 0 \quad \text{in } [A, B].$$

CHAPTER 3

Efficiency Within Nonlinear Thermoelasticity

19. The origins of thermodynamics lie in the attempt to understand, and exploit, the efficient conversion of heat energy into mechanical work by heat engines operating in a cycle. It is proposed to consider efficiency from the standpoint of each of our three theories in turn.

In the context of nonlinear thermoelasticity it will be supposed that the external force density and the external rate-of-heating density both vanish:

$$f = h = 0.$$

The same supposition will be in force in Chapter 5, which discusses efficiency from the standpoint of linearized thermoelasticity.

Within homogeneous and dissipationless thermoelasticity it is not possible to maintain the requirement $h = 0$ except in very special circumstances.

The decision to set $f = h = 0$ reflects the prejudice that it is probably unrealistic to grant ourselves the capacity to exert direct influence upon the interior of the body $[A, B]$. At best, we can hope to exert indirect influence by controlling conditions at the boundary points A and B—at which points we control, let us say, the displacement and the temperature.

In our treatment of nonlinear or linearized thermoelasticity, the body is heated either by the performance of work on it or by the conduction of heat across the boundary; it is not heated by adjustment of the external rate-of-heating density. Since heat conduction is absent from homogeneous and dissipationless thermoelasticity, adjustment of the external rate-of-heating density must take the place of heat conduction within that theory.

In what follows, the rate of working by the body on its exterior is

$$W = -[\sigma \dot{u}]_A^B,$$

and the net heat flux into the body from its exterior is

$$Q = [q]_A^B.$$

The net heat flux, which can be positive, negative, or zero, may be expressed as the difference

$$Q = Q^+ - Q^-$$

between the rate of absorption of heat by the body, which is

$$Q^+ = \text{Max}(Q, 0),$$

and the rate of emission of heat by the body, which is

$$Q^- = -\text{Min}(Q, 0).$$

Neither Q^+ nor Q^- can be negative.

Certain integrals figure prominently. These are:

$\int_{t_1}^{t_2} W \, dt =$ the work done by the body,
$\int_{t_1}^{t_2} Q \, dt =$ the net heat gained by the body,
$\int_{t_1}^{t_2} Q^+ \, dt =$ the heat absorbed by the body,
$\int_{t_1}^{t_2} Q^- \, dt =$ the heat emitted by the body.

By virtue of the relation $Q = Q^+ - Q^-$, the net heat gained by the body is the difference

$$\int_{t_1}^{t_2} Q \, dt = \int_{t_1}^{t_2} Q^+ \, dt - \int_{t_1}^{t_2} Q^- \, dt$$

between the heat absorbed by the body and the heat emitted by it.

The efficiency is the ratio

$$\int_{t_1}^{t_2} W \, dt \bigg/ \int_{t_1}^{t_2} Q^+ \, dt,$$

of the work done by the body to the heat absorbed by it, and is defined whenever the denominator is positive.

20. We are now in position to state our first result:

Suppose that

(i) $f = h = 0$ *on* $[A, B] \times [t_1, t_2]$,
(ii) $T(A, t) = T(B, t)$ $(= \tau(t)$ *say) on* $[t_1, t_2]$,
(iii) $[\frac{1}{2} \int_A^B \rho \dot{u}^2 \, dx + \int_A^B U \, dx]_{t_1}^{t_2} = 0$,
   $[\int_A^B S \, dx]_{t_1}^{t_2} = 0$.

Then

$$\int_{t_1}^{t_2} W \, dt = \int_{t_1}^{t_2} Q \, dt$$

$$\int_{t_1}^{t_2} \frac{Q}{\tau} \, dt + \int_{t_1}^{t_2} \int_A^B \frac{gq}{T^2} \, dx \, dt = 0,$$

$$\int_{t_1}^{t_2} W \, dt \leq \left(1 - \frac{m}{M}\right) \int_{t_1}^{t_2} Q^+ \, dt,$$

where M and m are, respectively, the maximum and minimum values attained by $\tau(t)$ *on* $[t_1, t_2]$.

The first hypothesis has been commented upon already.

The second envisages the body as being immersed in an environment whose temperature $\tau(t)$ is spatially homogeneous but variable in time.

The third requires the body to perform a cycle, in the sense that the sum of the kinetic energy and the internal energy returns, at the time t_2, to the value it had at the time t_1, and the same is true of the entropy. The hypothesis would be satisfied if, at each point x of $[A, B]$,

$$\dot{u}(x, t_1) = \dot{u}(x, t_2),$$

$$E(x, t_1) = E(x, t_2),$$

$$T(x, t_1) = T(x, t_2),$$

but these conditions are very much more restrictive than is (iii).

The first conclusion is that the work done by the body must coincide with the net heat gained by it.

The second conclusion is a step in the proof of the third, and major, conclusion, which says that the efficiency cannot exceed an upper bound

$$1 - \frac{m}{M},$$

depending only upon the range of temperatures over which $\tau(t)$ varies.

21. To begin the proof of what §20 asserts we return to the energy balance law (§5) and set $f = h = 0$ and $[a, b] = [A, B]$, thereby obtaining the equation

$$-W + Q = \tfrac{1}{2} \overline{\int_A^B \rho \dot{u}^2 \, dx + \int_A^B U \, dx}.$$

In view of (iii) it must be that

$$\int_{t_1}^{t_2} W \, dt = \int_{t_1}^{t_2} Q \, dt.$$

The next step is to set $h = 0$ in an equation derived in §9. This yields

$$\dot{S} = \left(\frac{q}{T}\right)' + \frac{gq}{T^2}.$$

On integrating with respect to x, we find that

$$\overline{\int_A^B S \, dx} = \left[\frac{q}{T}\right]_A^B + \int_A^B \frac{gq}{T^2} \, dx.$$

By virtue of (ii),

$$\left[\frac{q}{T}\right]_A^B = \frac{1}{\tau}[q]_A^B = \frac{Q}{\tau},$$

and hence

$$\overline{\int_A^B S \, dx} = \frac{Q}{\tau} + \int_A^B \frac{gq}{T^2} \, dx.$$

An integration with respect to t, and a further appeal to (iii), now yield the identity

$$\int_{t_1}^{t_2} \frac{Q}{\tau} \, dt + \int_{t_1}^{t_2} \int_A^B \frac{gq}{T^2} \, dx \, dt = 0.$$

In order to arrive at the efficiency estimate, we write

$$\int_{t_1}^{t_2} \frac{Q}{\tau}\, dt = \int_{t_1}^{t_2} \frac{(Q^+ - Q^-)}{\tau}\, dt$$

$$= \frac{1}{M} \int_{t_1}^{t_2} Q^+\, dt - \frac{1}{m} \int_{t_1}^{t_2} Q^-\, dt + \int_{t_1}^{t_2} \left(\frac{1}{\tau} - \frac{1}{M}\right) Q^+\, dt$$

$$+ \int_{t_1}^{t_2} \left(\frac{1}{m} - \frac{1}{\tau}\right) Q^-\, dt,$$

and then we substitute the difference

$$\int_{t_1}^{t_2} Q^+\, dt - \int_{t_1}^{t_2} W\, dt$$

for the term

$$\int_{t_1}^{t_2} Q^-\, dt,$$

and find that

$$\int_{t_1}^{t_2} \frac{Q}{\tau}\, dt = \frac{1}{m} \int_{t_1}^{t_2} W\, dt + \left(\frac{1}{M} - \frac{1}{m}\right) \int_{t_1}^{t_2} Q^+\, dt$$

$$+ \int_{t_1}^{t_2} \left(\frac{1}{\tau} - \frac{1}{M}\right) Q^+\, dt + \int_{t_1}^{t_2} \left(\frac{1}{m} - \frac{1}{\tau}\right) Q^-\, dt.$$

When this last expression is substituted into the identity already derived, and the resulting equation is rearranged, one arrives at the formula

$$\int_{t_1}^{t_2} W\, dt = \left(1 - \frac{m}{M}\right) \int_{t_1}^{t_2} Q^+\, dt - m \int_{t_1}^{t_2} \left(\frac{1}{\tau} - \frac{1}{M}\right) Q^+\, dt$$

$$- m \int_{t_1}^{t_2} \left(\frac{1}{m} - \frac{1}{\tau}\right) Q^-\, dt - m \int_{t_1}^{t_2} \int_A^B \frac{gq}{T^2}\, dx\, dt$$

for the work done by the body.

Each of the integrands

$$\left(\frac{1}{\tau} - \frac{1}{M}\right) Q^+, \quad \left(\frac{1}{m} - \frac{1}{\tau}\right) Q^-, \quad \frac{gq}{T^2}$$

is nonnegative, the last being so because of the heat conduction inequality. Since m is positive it must be that

$$\int_{t_1}^{t_2} W\,dt \leqq \left(1 - \frac{m}{M}\right) \int_{t_1}^{t_2} Q^+\,dt.$$

In short, the efficiency estimate is correct.

22. The argument establishes more than has been claimed. Indeed, within nonlinear thermoelasticity the upper bound on efficiency is unattainable: whenever the efficiency is defined it is strictly less than

$$1 - \frac{m}{M}.$$

If hypotheses (i), (ii), *and* (iii) *of §20 are in force, it is impossible that the inequality*

$$\int_{t_1}^{t_2} Q^+\,dt > 0$$

and the equation

$$\int_{t_1}^{t_2} W\,dt = \left(1 - \frac{m}{M}\right) \int_{t_1}^{t_2} Q^+\,dt$$

should both be satisfied.

For, if the equation were to hold it would follow that

$$\int_{t_1}^{t_2} \int_A^B \frac{gq}{T^2}\,dx\,dt = 0$$

and this and the heat conduction inequality would imply that the temperature gradient

$$g = 0 \quad \text{in} \quad [A, B] \times [t_1, t_2].$$

In view of a result of §10, it would follow that the heat flux

$$q = 0 \quad \text{in} \quad [A, B] \times [t_1, t_2].$$

Hence the net heat flux

$$Q = [q]_A^B = 0 \quad \text{in} \quad [t_1, t_2]$$

and, therefore,

$$Q^+ = 0 \quad \text{in} \quad [t_1, t_2]$$

and, finally,

$$\int_{t_1}^{t_2} Q^+ \, dt = 0.$$

Thus, the inequality and the equation cannot hold simultaneously.

23. Although the upper bound is not attainable within the nonlinear theory, there remains the possibility that it can be approximated arbitrarily closely and is, therefore, the least upper bound on efficiency.

Thus we are led to pose the question:

Let M, m, ε be positive numbers which satisfy

$$M > m$$

but are otherwise arbitrary. Does there exist an interval $[t_1, t_2]$, and do there exist fields u and T which are C^2 on $[A, B] \times [t_1, t_2]$ and satisfy the nonlinear displacement–temperature equations and, at the same time, ensure that

  (i) $f = h = 0$ on $[A, B] \times [t_1, t_2]$,
  (ii) $T(A, t) = T(B, t) \ (= \tau(t) \ say)$ on $[t_1, t_2]$,
  (iii) $\left[\frac{1}{2}\int_A^B \rho \dot{u}^2 \, dx + \int_A^B U \, dx\right]_{t_1}^{t_2} = 0$,
    $\left[\int_A^B S \, dx\right]_{t_1}^{t_2} = 0$,
  (iv) Max $\tau = M$, Min $\tau = m$,
  (v) $1 + E > 0$, $T > 0$ on $[A, B] \times [t_1, t_2]$,
  (vi) $\int_{t_1}^{t_2} Q^+ \, dt > 0$,
  (vii) $\int_{t_1}^{t_2} W \, dt > (1 - m/M - \varepsilon) \int_{t_1}^{t_2} Q^+ \, dt$?

The answer is almost certainly "yes", at least if the response of the body is suitably restricted, but I know no proof.

One factor that would have to be kept in mind in attempting to construct a proof is that the duration $t_2 - t_1$ of the underlying time interval would have to be suitably large, and the displacement and temperature fields would have to vary slowly with t—in accordance with the expectation that, in order to maximize efficiency, it is necessary to operate "quasi-statically".

It is possible, as Chapter 5 will show, to answer the corresponding question within linearized thermoelasticity, but even there the proof is lengthy. That chapter may point the way to the type of argument that will be required within the nonlinear theory, but there can be no easy passage from the linearized theory to the full nonlinear theory unless attention is confined to cases in which the difference $M - m$ is small by comparison with M.

The fundamental difficulty is to face up to the presence of the term

$$T\dot{S}$$

in the second reduced energy equation. It is just this term, though, which the linearized theory refuses to confront, for it seeks to replace that term with the term

$$T_0\dot{S}.$$

The question raised in this section serves to illustrate the wealth of investigation into thermodynamics that remains to be carried out; the ratio of established results to plausible speculations is regrettably meagre.

CHAPTER 4

Efficiency Within Homogeneous and Dissipationless Thermoelasticity

24. As has been noted, the requirement that the external rate-of-heating density $h = 0$, which was adopted in §19, cannot be maintained within homogeneous and dissipationless thermoelasticity.

In that theory, the rate of heating of the body by its exterior is

$$H = \int_A^B h \, dx = (B - A)h.$$

The change of symbol, with H replacing the Q of §19, is made in order to remind us that the mechanism by which the body is heated is not that of heat conduction.

The rate of heating of the body may be expressed as the difference

$$H = H^+ - H^-$$

between the rate of absorption of heat by the body, which is

$$H^+ = \text{Max}(H, 0),$$

and the rate of emission of heat by the body, which is

$$H^- = -\text{Min}(H, 0).$$

Neither H^+ nor H^- can be negative.

Since the stress is spatially homogeneous, the rate of working by the body on its exterior is

$$W = -[\sigma \dot{u}]_A^B = -\sigma [\dot{u}]_A^B,$$

where

$$[\dot{u}]_A^B = \int_A^B \dot{E}\, dx = (B - A)\dot{E},$$

and, hence,

$$W = -(B - A)\sigma\dot{E}.$$

Furthermore,

$\int_{t_1}^{t_2} W\, dt$ = the work done by the body,
$\int_{t_1}^{t_2} H\, dt$ = the net heat gained by the body,
$\int_{t_1}^{t_2} H^+\, dt$ = the heat absorbed by the body,
$\int_{t_1}^{t_2} H^-\, dt$ = the heat emitted by the body,

and

$$\int_{t_1}^{t_2} H\, dt = \int_{t_1}^{t_2} H^+\, dt - \int_{t_1}^{t_2} H^-\, dt.$$

The efficiency is the ratio

$$\int_{t_1}^{t_2} W\, dt \Big/ \int_{t_1}^{t_2} H^+\, dt.$$

The counterpart to the statement of §20 is:

Suppose that

$$[U]_{t_1}^{t_2} = [S]_{t_1}^{t_2} = 0.$$

Then

$$\int_{t_1}^{t_2} W\, dt = \int_{t_1}^{t_2} H\, dt,$$

$$\int_{t_1}^{t_2} \frac{H}{T}\, dt = 0,$$

$$\int_{t_1}^{t_2} W\, dt \le \left(1 - \frac{m}{M}\right) \int_{t_1}^{t_2} H^+\, dt,$$

where M and m are, respectively, the maximum and minimum values attained by $T(t)$ on $[t_1, t_2]$.

Three comments may help to draw the comparison with §20.

Firstly, the condition $f = 0$, which is a hypothesis in §20, is one of the simplifications presumed to be in force in the setting up of the homogeneous and dissipationless theory.

Secondly, hypothesis (ii) of §20 is automatically satisfied here because the temperature $T(t)$ is spatially homogeneous.

Thirdly, the kinetic energy, which occurs in (iii) of §20, is absent in the present context because inertia has been neglected.

25. The proof of the assertion of §24 is straightforward. Indeed, Gibbs's relation (§12) and the equation

$$h = T\dot{S}$$

imply that

$$\dot{U} = \sigma\dot{E} + h.$$

On integrating with respect to t, and invoking the hypothesis

$$[U]_{t_1}^{t_2} = 0,$$

we obtain

$$\int_{t_1}^{t_2} W\, dt = -(B - A) \int_{t_1}^{t_2} \sigma\dot{E}\, dt$$

$$= (B - A) \int_{t_1}^{t_2} (-\dot{U} + h)\, dt$$

$$= (B - A) \int_{t_1}^{t_2} h\, dt$$

$$= \int_{t_1}^{t_2} H\, dt.$$

Next, we note that

$$\int_{t_1}^{t_2} \frac{h}{T}\, dt = \int_{t_1}^{t_2} \dot{S}\, dt = [S]_{t_1}^{t_2} = 0$$

and, therefore,

$$\int_{t_1}^{t_2} \frac{H}{T}\, dt = 0.$$

The upper bound on efficiency can be established on much the same lines as before. What has just been proved yields the equations

$$0 = \int_{t_1}^{t_2} \frac{(H^+ - H^-)}{T} \, dt$$

$$= \frac{1}{M} \int_{t_1}^{t_2} H^+ \, dt - \frac{1}{m} \int_{t_1}^{t_2} H^- \, dt$$

$$+ \int_{t_1}^{t_2} \left(\frac{1}{T} - \frac{1}{M} \right) H^+ \, dt + \int_{t_1}^{t_2} \left(\frac{1}{m} - \frac{1}{T} \right) H^- \, dt,$$

and when we substitute

$$\int_{t_1}^{t_2} H^+ \, dt - \int_{t_1}^{t_2} W \, dt$$

for the term

$$\int_{t_1}^{t_2} H^- \, dt$$

and rearrange the resulting equation, we arrive at the formula

$$\int_{t_1}^{t_2} W \, dt = \left(1 - \frac{m}{M} \right) \int_{t_1}^{t_2} H^+ \, dt - m \int_{t_1}^{t_2} \left(\frac{1}{T} - \frac{1}{M} \right) H^+ \, dt$$

$$- m \int_{t_1}^{t_2} \left(\frac{1}{m} - \frac{1}{T} \right) H^- \, dt$$

for the work done by the body. Since each of the integrands

$$\left(\frac{1}{T} - \frac{1}{M} \right) H^+, \quad \left(\frac{1}{m} - \frac{1}{T} \right) H^-$$

is nonnegative, and m is positive, we have

$$\int_{t_1}^{t_2} W \, dt \leq \left(1 - \frac{m}{M} \right) \int_{t_1}^{t_2} H^+ \, dt$$

as required.

26. By contrast with nonlinear thermoelasticity, the neglect of heat conduction and inertia within homogeneous and dissipationless thermoelasticity ensures that the upper bound

$$1 - \frac{m}{M}$$

is attainable; that is to say, appropriate choices of the spatially homogeneous strain and temperature fields ensure that the efficiency is defined, and coincides with the upper bound.

In order for this to be true we need to be assured that, for given values of T and S, the equation

$$\hat{S}(E, T) = S$$

can be solved for E, at least locally.

It will be recalled that $\hat{S}$ is already subject to the requirement

$$\hat{S} = -\partial_T \hat{F},$$

where $\hat{F}$ is C^2. Hence $\hat{S}$ is C^1. The following additional restriction is postulated:

There are numbers E_0, T_0, S_0 such that

$$1 + E_0 > 0, \qquad T_0 > 0,$$

$$\hat{S}(E_0, T_0) = S_0, \qquad \partial_E \hat{S}(E_0, T_0) \neq 0.$$

In the light of Maxwell's relation, the last condition might equally well be replaced by

$$\partial_T \hat{\sigma}(E_0, T_0) \neq 0.$$

The temperature T_0 need have no connection with the reference temperature of the linearized theory.

The significance of the postulate is embodied in the following assertion:

There are numbers δ and ε which lie in the intervals

$$0 < \delta < 1 + E_0, \qquad 0 < \varepsilon < T_0,$$

and have the property that, to each T in $|T - T_0| < \varepsilon$ and each S in $|S - S_0| < \varepsilon$, there corresponds exactly one E in $|E - E_0| < \delta$ which satisfies

$$\hat{S}(E, T) = S.$$

The E whose existence is asserted will be written as

$$\check{E}(T, S).$$

Thus,

$$\hat{S}(\check{E}(T, S), T) = S.$$

It should be noted that the inequalities

$$\delta < 1 + E_0, \qquad |E - E_0| < \delta$$

imply that

$$1 + E > 1 + E_0 - \delta > 0$$

and, therefore,

$$1 + \check{E}(T, S) > 0.$$

Likewise, the inequalities

$$\varepsilon < T_0, \qquad |T - T_0| < \varepsilon$$

imply that

$$T > T_0 - \varepsilon > 0.$$

27. The assertion of the preceding section is proved by arguments of a familiar type but none the less the details will be set out for the sake of completeness.

I deal only with the case

$$\partial_E \hat{S}(E_0, T_0) > 0.$$

When the partial derivative is negative the obvious emendations are sufficient to carry the proof through.

Since $\partial_E \hat{S}$ is C^0, it is possible to choose δ in such a way that

$$0 < \delta < \mathrm{Min}(1 + E_0, T_0)$$

and to ensure at the same time that the inequality

$$\partial_E \hat{S}(E, T) > 0$$

holds throughout the open square

$$|E - E_0| < \delta, \qquad |T - T_0| < \delta$$

of the E, T-plane. Because of the restriction upon δ, points (E, T) of the open square satisfy

$$1 + E > 0, \qquad T > 0.$$

Moreover,

$$\hat{S}(E_0 + \delta, T_0) > \hat{S}(E_0, T_0), \qquad \hat{S}(E_0 - \delta, T_0) < \hat{S}(E_0, T_0),$$

or, in other words,

$$\hat{S}(E_0 + \delta, T_0) - S_0 > 0, \qquad \hat{S}(E_0 - \delta, T_0) - S_0 < 0.$$

In view of the fact that the functions

$$(T, S) \to \hat{S}(E_0 + \delta, T) - S,$$

$$(T, S) \to \hat{S}(E_0 - \delta, T) - S$$

are C^0 it must be possible to choose ε so that

$$0 < \varepsilon < \delta$$

and, at the same time, ensure that

$$\hat{S}(E_0 + \delta, T) - S > 0, \qquad \hat{S}(E_0 - \delta, T) - S < 0,$$

whenever

$$|T - T_0| < \varepsilon, \qquad |S - S_0| < \varepsilon.$$

Thus, if T and S satisfy these last two inequalities there must exist at least one E in

$$|E - E_0| < \delta$$

such that

$$\hat{S}(E, T) = S.$$

If there were more than one such E it would follow that

$$\partial_E \hat{S}(E_1, T_1) = 0$$

for some E_1 and some T_1 satisfying

$$|E_1 - E_0| < \delta, \qquad |T_1 - T_0| < \varepsilon.$$

Since $\varepsilon < \delta$ the vanishing of $\partial_E \hat{S}$ is ruled out by the choice of δ and, therefore, E is unique.

28. It is now possible to show that the upper bound on efficiency is attainable.

Let the postulate of §26 be in force, let δ and ε be as in §27, let M and m be any numbers which satisfy

$$T_0 - \varepsilon < m < M < T_0 + \varepsilon,$$

and let $[t_1, t_2]$ be any closed and bounded interval. Then there are homogeneous strain and temperature fields, $E(t)$ and $T(t)$, which are defined on $[t_1, t_2]$ and which ensure that

(i) $[U]_{t_1}^{t_2} = [S]_{t_1}^{t_2} = 0$,

(ii) Max $T = M$, Min $T = m$,

(iii) $1 + E > 0$ and $T > 0$ in $[t_1, t_2]$,

(iv) $\int_{t_1}^{t_2} H^+ \, dt > 0$,

(v) $\int_{t_1}^{t_2} W \, dt = (1 - m/M) \int_{t_1}^{t_2} H^+ \, dt$.

It should be noted that, within homogeneous and dissipationless thermoelasticity, no restriction is required upon the duration of the interval $[t_1, t_2]$. By contrast, theories which take account of heat conduction and inertia require the duration to be large—as the arguments of Chapter 5 will illustrate.

29. To prove what §28 asserts to be the case, let

$$t_1 < s_1 < s_2 < s_3 < s_4 < t_2$$

be a partition of the interval $[t_1, t_2]$, and let $T(t)$ be chosen so that

$$[T]_{t_1}^{t_2} = 0,$$

$$m \leq T(t) \leq M \quad \text{in } [t_1, t_2],$$

$$T(t) = M \quad \text{in } [s_1, s_2],$$

$$T(t) = m \quad \text{in } [s_3, s_4].$$

Furthermore, let $\phi(t)$ and $\psi(t)$ be chosen to satisfy the conditions

$$\phi(t) = 0 \quad \text{in } [t_1, s_1] \cup [s_2, t_2],$$

$$\phi(t) > 0 \quad \text{in } (s_1, s_2),$$

$$\psi(t) = 0 \quad \text{in } [t_1, s_3] \cup [s_4, t_2],$$

$$\psi(t) > 0 \quad \text{in } (s_3, s_4),$$

$$\int_{s_1}^{s_2} \phi(t) \, dt = \int_{s_3}^{s_4} \psi(t) \, dt.$$

(Note that it is possible to arrange that each of T, ϕ, ψ is C^∞.)

Next, put

$$h(t) = \begin{cases} 0 & \text{in } [t_1, s_1] \cup [s_2, s_3] \cup [s_4, t_2], \\ \alpha M \phi(t) & \text{in } [s_1, s_2], \\ -\alpha m \psi(t) & \text{in } [s_3, s_4], \end{cases}$$

and

$$S(t) = S_0 + \int_{t_1}^{t} \frac{h(s)}{T(s)} \, ds,$$

where α is a positive constant at our disposal.

By choosing α sufficiently small, we can arrange that

$$|S(t) - S_0| < \varepsilon \quad \text{in } [t_1, t_2].$$

Moreover, the choice of T ensures that

$$T_0 - \varepsilon < m \leq T(t) \leq M < T_0 + \varepsilon \quad \text{in } [t_1, t_2]$$

and, hence, that

$$|T(t) - T_0| < \varepsilon \quad \text{in } [t_1, t_2].$$

Thus §26 permits us to define the homogeneous strain

$$E(t) = \check{E}(T(t), S(t))$$

and this definition ensures that $1 + E > 0$. It is, of course, the case that $T > 0$.

Furthermore, the definitions of S and h ensure that

$$h = T\dot{S}$$

and that

$$\int_{t_1}^{t_2} \frac{h}{T} \, dt = \left(\int_{s_1}^{s_2} + \int_{s_3}^{s_4} \right) \frac{h}{T} \, dt$$

$$= \alpha \int_{s_1}^{s_2} \phi \, dt - \alpha \int_{s_3}^{s_4} \psi \, dt$$

$$= 0.$$

Hence,

$$[S]_{t_1}^{t_2} = \int_{t_1}^{t_2} \dot{S} \, dt = \int_{t_1}^{t_2} \frac{h}{T} \, dt = 0.$$

In addition,

$$E(t_1) = \check{E}(T(t_1), S(t_1)) = \check{E}(T(t_2), S(t_2)) = E(t_2)$$

and, therefore,

$$U(t_1) = \hat{U}(E(t_1), T(t_1)) = \hat{U}(E(t_2), T(t_2)) = U(t_2),$$

that is to say

$$[U]_{t_1}^{t_2} = 0.$$

The rate of heating of the body $H(t)$ is

$$\begin{cases} 0 & \text{in } [t_1, s_1] \cup [s_2, s_3] \cup [s_4, t_2], \\ > 0 & \text{in } (s_1, s_2), \\ < 0 & \text{in } (s_3, s_4), \end{cases}$$

and so the heat absorbed by the body is

$$\int_{t_1}^{t_2} H^+ \, dt = (B - A)\alpha M \int_{s_1}^{s_2} \phi \, dt,$$

which is strictly positive.

Finally, we check that the efficiency attains the value

$$1 - \frac{m}{M}$$

by returning to a formula for the work done by the body which was derived in §25, namely

$$\int_{t_1}^{t_2} W \, dt = \left(1 - \frac{m}{M}\right) \int_{t_1}^{t_2} H^+ \, dt - m \int_{t_1}^{t_2} \left(\frac{1}{T} - \frac{1}{M}\right) H^+ \, dt$$
$$- m \int_{t_1}^{t_2} \left(\frac{1}{m} - \frac{1}{T}\right) H^- \, dt.$$

Since $T(t) = M$ in $[s_1, s_2]$, it must be that

$$\int_{t_1}^{t_2} \left(\frac{1}{T} - \frac{1}{M}\right) H^+ \, dt = \int_{s_1}^{s_2} \left(\frac{1}{T} - \frac{1}{M}\right) H^+ \, dt = 0,$$

and, since $T(t) = m$ in $[s_3, s_4]$, that

$$\int_{t_1}^{t_2} \left(\frac{1}{m} - \frac{1}{T}\right) H^- \, dt = \int_{s_3}^{s_4} \left(\frac{1}{m} - \frac{1}{T}\right) H^- \, dt = 0,$$

and, therefore,

$$\int_{t_1}^{t_2} W \, dt = \left(1 - \frac{m}{M}\right) \int_{t_1}^{t_2} H^+ \, dt$$

as required.

What has been constructed in the course of the proof is a Carnot cycle on the interval $[t_1, t_2]$, by which is meant a pair of spatially homogeneous fields $(E(t), T(t))$ which satisfy

(i) $[E]_{t_1}^{t_2} = [T]_{t_1}^{t_2} = 0,$
(ii) $T(t) = M$ at every t at which $H(t) > 0,$
(iii) $T(t) = m$ at every t at which $H(t) < 0,$

M and m being, respectively, the maximum and minimum values attained by T on $[t_1, t_2]$.

According to (ii) and (iii), heat is absorbed only at the maximum temperature at which the Carnot cycle operates, and heat is emitted only at the minimum temperature at which the Carnot cycle operates.

30. As has been pointed out (§22), the upper bound on efficiency is not attainable within nonlinear thermoelasticity, and, within that theory, it is an important open problem to show that the bound is best possible in the sense that it can be approached arbitrarily closely. That being so, it is of interest to examine an argument which, in the context of homogeneous and dissipationless thermoelasticity, establishes less than does §29, but, none the less, suffices to show that the bound

$$1 - \frac{m}{M}$$

cannot be replaced by any smaller number. A related, but more difficult, line of argument proves to be effective in linearized thermoelasticity.

Let $E_0, T_0, S_0, \delta, \varepsilon$ be as in §26, let M and m be any numbers which satisfy

$$T_0 - \varepsilon < m < M < T_0 + \varepsilon,$$

and let $[t_1, t_2]$ be any closed and bounded interval. Then there is a sequence $\{(E_n, T_n)\}_{n \geq 1}$ of pairs of homogeneous strain and temperature fields, which are defined on $[t_1, t_2]$ and satisfy

(i) $[U_n]_{t_1}^{t_2} = [S_n]_{t_1}^{t_2} = 0,$
(ii) Max $T_n = M$, Min $T_n = m$,
(iii) $1 + E_n > 0$ *and* $T_n > 0$ *on* $[t_1, t_2]$,
(iv) $\int_{t_1}^{t_2} H_n^+ \, dt > 0,$
(v) $\int_{t_1}^{t_2} W_n \, dt / \int_{t_1}^{t_2} H_n^+ \, dt \to 1 - m/M$ *as* $n \to \infty$.

In this statement, U_n, S_n, H_n^+, W_n are, respectively, the internal energy density, the entropy density, the rate of absorption of heat by the body, and the rate of working by the body that correspond to the homogeneous strain field E_n and the homogeneous temperature field T_n.

In order to construct the sequence it is convenient to introduce the integrals

$$I_n = \int_0^{\pi/2} \cos^n(s)\, ds = \begin{cases} \dfrac{2^{n-1}[((n-1)/2)!]^2}{n!}, & n \text{ odd}, \\[4mm] \dfrac{\pi n!}{2^{n+1}[(n/2)!]^2}, & n \text{ even}. \end{cases}$$

Since

$$I_{n+1} < I_n < I_{n-1}$$

and since, as an integration by parts reveals,

$$I_{n-1} = \left(\frac{n+1}{n}\right) I_{n+1},$$

it must be that

$$\frac{n}{n+1} < \frac{I_{n+1}}{I_n} < 1$$

and, hence,

$$I_{n+1}/I_n \to 1 \qquad \text{as} \quad n \to \infty.$$

Stirling's asymptotic evaluation of the factorials leads to the same conclusion by a less elementary route.

Each of the temperature fields is chosen in the same way, that is as

$$T_n(t) = \tfrac{1}{2}(M + m) + \tfrac{1}{2}(M - m) \cos\left\{\frac{2\pi(t - t_1)}{t_2 - t_1}\right\}.$$

Hence,

$$[T_n]_{t_1}^{t_2} = 0, \qquad \text{Max } T_n = M, \qquad \text{Min } T_n = m, \qquad T_n > 0,$$

and, because of the inequalities

$$T_0 - \varepsilon < m < M < T_0 + \varepsilon,$$

it must be that

$$|T_n(t) - T_0| < \varepsilon \quad \text{in } [t_1, t_2].$$

The entropy density is chosen to be

$$S_n(t) = S_0 + \alpha \int_{t_1}^{t} \cos^{2n+1} \left\{ \frac{2\pi(s - t_1)}{t_2 - t_1} \right\} ds,$$

where α meets the restrictions $0 < \alpha(t_2 - t_1) < \varepsilon$. The restrictions ensure that

$$|S_n(t) - S_0| < \varepsilon \quad \text{in } [t_1, t_2].$$

It is now permissible to define the strain fields

$$E_n(t) = \check{E}(T_n(t), S_n(t)),$$

and these satisfy

$$1 + E_n > 0.$$

The fact that

$$\int_{t_1}^{t_2} \cos^{2n+1} \left\{ \frac{2\pi(t - t_1)}{t_2 - t_1} \right\} dt = 0$$

ensures that $S_n(t_2) = S_0 = S_n(t_1)$ and, therefore, that

$$[S_n]_{t_1}^{t_2} = 0,$$

$$E_n(t_2) = \check{E}(T_n(t_2), S_n(t_2)) = \check{E}(T_n(t_1), S_n(t_1)) = E_n(t_1),$$

and

$$U_n(t_2) = \hat{U}(E_n(t_2), T_n(t_2)) = \hat{U}(E_n(t_1), T_n(t_1)) = U_n(t_1).$$

That is to say

$$[U_n]_{t_1}^{t_2} = 0.$$

Thus (i), (ii), and (iii) are satisfied. It remains to check (iv) and (v). Because

$$\dot{S}_n(t) = \alpha \cos^{2n+1} \left\{ \frac{2\pi(t - t_1)}{t_2 - t_1} \right\}$$

the set of times at which $\dot{S}_n$ is positive is the union of intervals

$$[t_1, \tfrac{1}{4}(3t_1 + t_2)) \cup (\tfrac{1}{4}(t_1 + 3t_2), t_2]$$

and, because T_n is positive, the set of times at which the rate of heating of the body

$$H_n = (B - A)T_n \dot{S}_n$$

is positive is the same union of intervals. Hence, the heat absorbed by the body is

$$\int_{t_1}^{t_2} H_n^+ \, dt = \left(\int_{t_1}^{(3t_1+t_2)/4} + \int_{(t_1+3t_2)/4}^{t_2} \right) (B - A) T_n \dot{S}_n \, dt$$

$$= (B - A)\alpha \left(\int_{t_1}^{(3t_1+t_2)/4} + \int_{(t_1+3t_2)/4}^{t_2} \right)$$

$$\left[\tfrac{1}{2}(M + m) \cos^{2n+1}\left\{ \frac{2\pi(t - t_1)}{t_2 - t_1} \right\} \right.$$

$$\left. + \tfrac{1}{2}(M - m) \cos^{2n+2}\left\{ \frac{2\pi(t - t_1)}{t_2 - t_1} \right\} \right] dt$$

$$= (B - A)\alpha \int_{t_1}^{(3t_1+t_2)/4} \left[(M + m) \cos^{2n+1}\left\{ \frac{2\pi(t - t_1)}{t_2 - t_1} \right\} \right.$$

$$\left. + (M - m) \cos^{2n+2}\left\{ \frac{2\pi(t - t_1)}{t_2 - t_1} \right\} \right] dt$$

and, on introducing the change of variable

$$s = \frac{2\pi(t - t_1)}{t_2 - t_1},$$

we find that

$$\int_{t_1}^{t_2} H_n^+ \, dt = \frac{(B - A)\alpha(t_2 - t_1)}{2\pi}((M + m)I_{2n+1} + (M - m)I_{2n+2}) > 0.$$

Thus (iv) holds.

Furthermore, the work done by the body coincides with the net heat gained by the body (§24). Thus

$$\int_{t_1}^{t_2} W_n \, dt = \int_{t_1}^{t_2} H_n \, dt$$

$$= \int_{t_1}^{t_2} (B - A) T_n \dot{S}_n \, dt$$

$$= (B - A)\alpha \int_{t_1}^{t_2} \left[\tfrac{1}{2}(M + m) \cos^{2n+1}\left\{ \frac{2\pi(t - t_1)}{t_2 - t_1} \right\} \right.$$

$$\left. + \tfrac{1}{2}(M - m) \cos^{2n+2}\left\{ \frac{2\pi(t - t_1)}{t_2 - t_1} \right\} \right] dt$$

$$= \tfrac{1}{2}(B - A)\alpha(M - m) \int_{t_1}^{t_2} \cos^{2n+2}\left\{\frac{2\pi(t - t_1)}{t_2 - t_1}\right\} dt$$

$$= \frac{(B - A)\alpha(t_2 - t_1)}{\pi}(M - m)I_{2n+2}$$

and the efficiency is

$$\frac{\int_{t_1}^{t_2} W_n \, dt}{\int_{t_1}^{t_2} H_n^+ \, dt} = \frac{2(M - m)}{(M + m)(I_{2n+1}/I_{2n+2}) + M - m}.$$

When $n \to \infty$ this ratio converges to

$$\frac{2(M - m)}{M + m + M - m} = 1 - \frac{m}{M}$$

and, hence, (v) is correct.

It is not difficult to see why the construction should be effective. Indeed, $T_n(t)$ attains its maximum value M at the endpoints t_1 and t_2 of the interval $[t_1, t_2]$, and attains its minimum value m at the midpoint $\tfrac{1}{2}(t_1 + t_2)$. On the other hand, when n is large, the bulk of the positive part of the graph of $H_n(t)$ is concentrated near the endpoints, while the bulk of the negative part is concentrated near the midpoint. Thus, most of the heat absorbed by the body is absorbed at temperatures close to the maximum, and most of the heat emitted by the body is emitted at temperatures close to the minimum; in brief, the pair $(E_n(t), T_n(t))$ is approximately a Carnot cycle when n is large.

CHAPTER 5

Efficiency Within Linearized Thermoelasticity

31. I turn now to the linearized theory set out in Part 2 of Chapter 2.

It will be found that the approximations involved in the linearization, and especially the replacement of the term

$$T\dot{S}$$

in the second reduced energy equation by the term

$$T_0\dot{S},$$

distort the truth to the extent of predicting an upper bound on efficiency which differs slightly from that predicted by nonlinear thermoelasticity and by homogeneous and dissipationless thermoelasticity. On the other hand, it transpires that, while the bound is unattainable, it is none the less the best possible bound within the linearized theory, for it can be approached arbitrarily closely—as we shall demonstrate.

As in Chapter 3, and for the same reasons, it is supposed that the external force density and the external rate-of-heating density both vanish, that is to say

$$f = h = 0.$$

Certain of the definitions of Chapter 3 remain in force. Thus, the rate of working by the body on its exterior is

$$W = -[\sigma\dot{u}]_A^B.$$

The net heat flux into the body from its exterior is

$$Q = [q]_A^B,$$

and this is expressible as the difference

$$Q = Q^+ - Q^-$$

between the rate of absorption of heat by the body, which is

$$Q^+ = \text{Max}(Q, 0),$$

and the rate of emission of heat by the body, which is

$$Q^- = -\text{Min}(Q, 0).$$

Furthermore,

$\int_{t_1}^{t_2} W \, dt$ = the work done by the body,
$\int_{t_1}^{t_2} Q \, dt$ = the net heat gained by the body,
$\int_{t_1}^{t_2} Q^+ \, dt$ = the heat absorbed by the body,
$\int_{t_1}^{t_2} Q^- \, dt$ = the heat emitted by the body,

$$\int_{t_1}^{t_2} Q \, dt = \int_{t_1}^{t_2} Q^+ \, dt - \int_{t_1}^{t_2} Q^- \, dt,$$

and the efficiency is the ratio

$$\int_{t_1}^{t_2} W \, dt \bigg/ \int_{t_1}^{t_2} Q^+ \, dt$$

of the work done by the body to the heat absorbed by it.

32. What the linearized theory predicts is that the upper bound

$$1 - \frac{m}{M} = \frac{M - m}{M}$$

of the nonlinear theory, and of the homogeneous and dissipationless theory, is to be replaced by

$$\frac{M - m}{T_0}.$$

This is the main conclusion of the following counterpart to the result of §20:

Suppose that

(i) $f = h = 0$ *on* $[A, B] \times [t_1, t_2]$,
(ii) $T(A, t) = T(B, t) \ (= \tau(t) \ say)$ *on* $[t_1, t_2]$,

(iii) $[\frac{1}{2}\int_A^B \rho \dot{u}^2 \, dx + \int_A^B U \, dx]_{t_1}^{t_2} = 0,$
$[\int_A^B S \, dx]_{t_1}^{t_2} = 0.$

Then

$$\int_{t_1}^{t_2} Q \, dt = 0,$$

$$\int_{t_1}^{t_2} W \, dt + \frac{1}{T_0} \int_{t_1}^{t_2} \int_A^B gq \, dx \, dt = \frac{1}{T_0} \int_{t_1}^{t_2} \tau Q \, dt,$$

$$\int_{t_1}^{t_2} W \, dt \leq \frac{(M - m)}{T_0} \int_{t_1}^{t_2} Q^+ \, dt,$$

where M and m are, respectively, the maximum and minimum values attained by $\tau(t)$ on $[t_1, t_2]$.

The first conclusion, namely that the net heat gained by the body is zero, or equivalently that the heat absorbed by the body and the heat emitted by the body must coincide, is a surprising one for, it will be recalled, both nonlinear thermoelasticity and homogeneous and dissipationless thermoelasticity imply that the net heat gained by the body coincides with the work done by the body and does not generally vanish.

It is interesting to observe, though, that the equality of the heat absorbed and the heat emitted in a cycle was taken as axiomatic by Carnot in his memoir "Réflexions Sur La Puissance Motrice Du Feu" [2, p. 19]. Lord Kelvin drew attention to this assumption in his account of Carnot's theory [6, pp. 115–118].

33. In order to prove what §32 asserts we begin by observing that, since $h = 0$, the approximate second reduced energy equation becomes

$$q' = T_0 \dot{S},$$

and an integration with respect to x over $[A, B]$ yields the expression

$$Q = T_0 \overline{\int_A^B S \, dx}$$

for the net heat flux into the body. Because of (iii), a further integration with respect to t gives

$$\int_{t_1}^{t_2} Q \, dt = 0,$$

as required.

The second identity depends upon Gibbs's relation

$$\dot{U} = \sigma \dot{E} + T \dot{S}$$

which is known to be valid within linearized thermoelasticity (§17). Adding the term

$$\rho \dot{u} \ddot{u}$$

to each side, and taking the equation of motion

$$\sigma' = \rho \ddot{u}$$

into account, now yields

$$\rho \dot{u} \ddot{u} + \dot{U} = (\sigma \dot{u})' + T \dot{S}.$$

Next, we substitute q'/T_0 for $\dot{S}$ and find

$$\rho \dot{u} \ddot{u} + \dot{U} = (\sigma \dot{u})' + \frac{T}{T_0} q' = (\sigma \dot{u})' + \frac{1}{T_0} (Tq)' - \frac{gq}{T_0}.$$

On integrating with respect to x, and noting that

$$\int_A^B (\sigma \dot{u})' \, dx = [\sigma \dot{u}]_A^B = -W$$

and, because of (ii), that

$$\int_A^B (Tq)' \, dx = [Tq]_A^B = \tau [q]_A^B = \tau Q,$$

we obtain

$$\frac{1}{2} \overline{\int_A^B \rho \dot{u}^2 \, dx + \int_A^B U \, dx} = -W + \frac{1}{T_0} \tau Q - \frac{1}{T_0} \int_A^B gq \, dx.$$

Thus, an integration with respect to t and a further appeal to (iii) yield the identity

$$\int_{t_1}^{t_2} W \, dt + \frac{1}{T_0} \int_{t_1}^{t_2} \int_A^B gq \, dx \, dt = \frac{1}{T_0} \int_{t_1}^{t_2} \tau Q \, dt.$$

The estimate on efficiency can be deduced from this identity. For the integral on the right-hand side can be rewritten as

$$\int_{t_1}^{t_2} \tau Q \, dt = \int_{t_1}^{t_2} \tau Q^+ \, dt - \int_{t_1}^{t_2} \tau Q^- \, dt$$

$$= M \int_{t_1}^{t_2} Q^+ \, dt - m \int_{t_1}^{t_2} Q^- \, dt - \int_{t_1}^{t_2} (M - \tau) Q^+ \, dt$$

$$- \int_{t_1}^{t_2} (\tau - m) Q^- \, dt$$

$$= (M - m) \int_{t_1}^{t_2} Q^+ \, dt - \int_{t_1}^{t_2} (M - \tau) Q^+ \, dt$$

$$- \int_{t_1}^{t_2} (\tau - m) Q^- \, dt,$$

where we have used the equality

$$\int_{t_1}^{t_2} Q^+ \, dt = \int_{t_1}^{t_2} Q^- \, dt$$

of the heat absorbed by the body and the heat emitted by it. Thus, we obtain the formula

$$\int_{t_1}^{t_2} W \, dt = \frac{(M - m)}{T_0} \int_{t_1}^{t_2} Q^+ \, dt - \int_{t_1}^{t_2} (M - \tau) Q^+ \, dt$$

$$- \int_{t_1}^{t_2} (\tau - m) Q^- \, dt - \frac{1}{T_0} \int_{t_1}^{t_2} \int_A^B gq \, dx \, dt$$

for the work done by the body. Since each of the integrands

$$(M - \tau) Q^+, \quad (\tau - m) Q^-, \quad gq$$

is nonnegative (the last by virtue of the postulate that the thermal conductivity $k(x)$ is positive on $[A, B]$) it must be that

$$\int_{t_1}^{t_2} W \, dt \leqq \frac{(M - m)}{T_0} \int_{t_1}^{t_2} Q^+ \, dt.$$

34. The same argument shows that the bound on efficiency is unattainable.

If hypotheses (i), (ii), *and* (iii) *of* §32 *are in force, it is impossible that the inequality*

$$\int_{t_1}^{t_2} Q^+ \, dt > 0$$

and the equation

$$\int_{t_1}^{t_2} W\,dt = \frac{(M-m)}{T_0} \int_{t_1}^{t_2} Q^+\,dt$$

should both be satisfied.

For, if the equation were satisfied, it would follow that

$$\int_{t_1}^{t_2} \int_A^B gq\,dx\,dt = 0$$

and, as the thermal conductivity is positive, that

$$g = q = 0 \quad \text{in} \quad [A, B] \times [t_1, t_2].$$

Hence

$$Q = [q]_A^B = 0 \quad \text{in } [t_1, t_2],$$
$$Q^+ = 0 \quad \text{in } [t_1, t_2],$$

and

$$\int_{t_1}^{t_2} Q^+\,dt = 0,$$

and, therefore, the inequality cannot be correct in these circumstances.

35. In the context of linearized thermoelasticity it is possible to answer affirmatively the question which corresponds to that posed in §23. In order to do so it is necessary to restrict the coefficients that appear in the linearized displacement–temperature equations; the thermal conductivity k has already been assumed to be positive but we shall now require β, μ, ρ, c to be positive as well. The positivity of β, ρ, c requires no comment. In view of the constitutive relation

$$\sigma = \sigma_0 + \beta E - \mu(T - T_0),$$

the positivity of μ (and β) means that, at constant stress, the strain is an increasing function of the temperature–which would appear to be a reasonable requirement.

Suppose that $\beta(x), \mu(x), \rho(x), k(x), c(x)$ are positive and C^0 in $[A, B]$ and, in addition, that $\mu(x)$ is C^1. Let M, m, ε be positive numbers that satisfy

$$M > m$$

but are otherwise arbitrary. Then there exists an interval $[t_1, t_2]$, *and there exist fields* u *and* T *which are* C^2 *on* $[A, B] \times [t_1, t_2]$ *and satisfy the linearized displacement–temperature equations and, at the same time, ensure that*

(i) $f = h = 0$ *on* $[A, B] \times [t_1, t_2]$,
(ii) $T(A, t) = T(B, t) \,(= \tau(t)$ *say) on* $[t_1, t_2]$,
(iii) $[\frac{1}{2} \int_A^B \rho \dot{u}^2 \, dx + \int_A^B U \, dx]_{t_1}^{t_2} = 0$,
    $[\int_A^B S \, dx]_{t_1}^{t_2} = 0$,
(iv) Max $\tau = M$, Min $\tau = m$,
(v) $\int_{t_1}^{t_2} Q^+ \, dt > 0$,
(vi) $\int_{t_1}^{t_2} W \, dt > ((M - m)/T_0 - \varepsilon) \int_{t_1}^{t_2} Q^+ \, dt$.

The proof is lengthy and will occupy the remainder of the chapter. A number of subsidiary results have to be proved along the way.

36. Our first task is to face up to questions of existence and to be convinced that there is a sufficient supply of trigonometric solutions of the linearized displacement–temperature equations that can be maintained by controlling conditions at the boundary points A and B, and without supplying a nonzero external force density f or a nonzero rate-of-heating density h.

Let ω *be any real number, and let* $\phi_A, \phi_B, \theta_A, \theta_B$ *be any complex numbers. Then there are uniquely defined complex-valued functions* $\Phi(x)$ *and* $\Theta(x)$ *which satisfy the ordinary differential equations*

$$[\beta\Phi' - \mu\Theta]' = -\omega^2 \rho\Phi,$$

$$[k\Theta']' = i\omega T_0 \mu\Phi' + i\omega c\Theta,$$

and the boundary conditions

$$\Phi(A) = \phi_A, \qquad \Phi(B) = \phi_B, \qquad \Theta(A) = \theta_A, \qquad \Theta(B) = \theta_B.$$

In circumstances in which it is necessary to emphasize dependence upon all the parameters $\omega, \phi_A, \phi_B, \theta_A, \theta_B$ I shall use the extended notation

$$\Phi(x, \omega, \phi_A, \phi_B, \theta_A, \theta_B),$$

$$\Theta(x, \omega, \phi_A, \phi_B, \theta_A, \theta_B),$$

or, in circumstances in which it is the dependence upon ω that is of interest, the notation

$$\Phi(x, \omega), \quad \Theta(x, \omega).$$

The significance of what has been asserted lies in the fact that the real-valued functions

$$u(x, t) = \text{Re}[\Phi(x) \exp(i\omega t)],$$

$$T(x, t) = T_0 + \text{Re}[\Theta(x) \exp(i\omega t)]$$

are displacement and temperature fields which satisfy the linearized displacement–temperature equations, with $f = h = 0$, and are periodic in their dependence upon t, with period $2\pi/\omega$, that is to say

$$u(x, t) = u\left(x, t + \frac{2\pi}{\omega}\right), \qquad T(x, t) = T\left(x, t + \frac{2\pi}{\omega}\right).$$

The assertion of this section should be contrasted with the predictions of the isothermal theory of linearized elasticity. According to that theory the displacement is a solution of the partial differential equation

$$[\beta u']' = \rho \ddot{u},$$

which may be obtained from the first of the linearized displacement–temperature equations by setting $\mu = 0$. If we look for solutions of the form

$$u(x, t) = \text{Re}[\Phi(x) \exp(i\omega t)],$$

we are led to consider the boundary value problem

$$[\beta \Phi']' = -\omega^2 \rho \Phi, \qquad \Phi(A) = \phi_A, \qquad \Phi(B) = \phi_B.$$

It is no longer the case that Φ always exists, and nor need it be unique if it exists for the Sturm–Liouville problem which corresponds to the choices $\phi_A = \phi_B = 0$ will have a solution other than $\Phi(x) = 0$ whenever $\omega/2\pi$ is a characteristic frequency of vibration.

This last remark makes it plain that our results depend crucially upon the restriction $\mu > 0$ and upon the consequent coupling between the two linearized displacement–temperature equations.

37. The uniqueness of Φ and Θ is readily established. It is enough to show that

$$\Phi(x) = \Theta(x) = 0 \quad \text{in } [A, B]$$

if

$$\phi_A = \phi_B = \theta_A = \theta_B = 0.$$

The first differential equation implies that

$$[\beta\bar{\Phi}' - \mu\bar{\Theta}]' = -\omega^2\rho\bar{\Phi},$$

the bars denoting complex conjugates. On multiplying through by Φ, and rearranging the resulting expression, we find

$$[\Phi(\beta\bar{\Phi}' - \mu\bar{\Theta})]' = \beta|\Phi'|^2 - \mu\Phi'\bar{\Theta} - \omega^2\rho|\Phi|^2,$$

and if we integrate with respect to x and remember that Φ vanishes at the boundary points we see that the integral

$$\int_A^B \mu\Phi'\bar{\Theta}\, dx = \int_A^B (\beta|\Phi'|^2 - \omega^2\rho|\Phi|^2)\, dx$$

and is, therefore, real-valued.

Next, we multiply the second differential equation through by $\bar{\Theta}$, rearrange the resulting expression, and obtain

$$[\bar{\Theta}k\Theta']' = k|\Theta'|^2 + i\omega T_0\mu\Phi'\bar{\Theta} + i\omega c|\Theta|^2.$$

Integration with respect to x, and use of the fact that Θ vanishes at the boundary points, now yield

$$\int_A^B k|\Theta'|^2\, dx + i\omega T_0 \int_A^B \mu\Phi'\bar{\Theta}\, dx + i\omega \int_A^B c|\Theta|^2\, dx = 0$$

and, when we substitute for the second of these integrals, we obtain

$$\int_A^B k|\Theta'|^2\, dx + i\omega \int_A^B [T_0(\beta|\Phi'|^2 - \omega^2\rho|\Phi|^2) + c|\Theta|^2]\, dx = 0.$$

The first term on the left-hand side is real, while the second is a pure imaginary number. Hence

$$\int_A^B k|\Theta'|^2\, dx = 0.$$

Since the thermal conductivity $k(x)$ is positive, it must be that $\Theta'(x) = 0$ and, because Θ vanishes at the boundary points, that $\Theta(x) = 0$.

That being so, the second differential equation reduces to

$$i\omega T_0\mu\Phi' = 0.$$

Since T_0 and μ are positive, it follows that $\Phi'(x) = 0$ provided $\omega \neq 0$.

Because Φ vanishes at the boundary points, it must be that $\Phi(x) = 0$ and, hence, the proof of uniqueness is complete as long as $\omega \neq 0$.

When $\omega = 0$ the proof that $\Theta(x) = 0$ goes through exactly as before. In order to conclude that $\Phi(x) = 0$, we return to the first differential equation and find

$$[\beta\Phi']' = 0.$$

Hence $\beta(x)\Phi'(x)$ is a constant, κ say. Since β is positive and Φ vanishes at the boundary points

$$0 = \int_A^B \Phi' \, dx = \kappa \int_A^B \frac{1}{\beta} \, dx.$$

Thus $\kappa = 0$, $\Phi'(x) = 0$ and, finally, $\Phi(x) = 0$. Hence the proof of uniqueness is complete.

38. The proof of existence depends upon showing that the boundary value problem of §36 is equivalent to a certain integral equation of the Fredholm type.

We begin by considering the boundary value problem

$$[k\Theta']' = i\omega T_0\mu\Phi' + i\omega c\Theta,$$

$$\Theta(A) = \theta_A, \qquad \Theta(B) = \theta_B,$$

which is to be regarded, for the moment, as a problem to be solved for Θ, given that the term $i\omega T_0\mu\Phi'$ is known.

First, we note that the homogeneous problem

$$[k\xi']' = i\omega c\xi, \qquad \xi(A) = \xi(B) = 0,$$

has no complex-valued solution other than $\xi(x) = 0$.

For, on multiplying the differential equation through by the complex conjugate $\bar{\xi}$, we obtain

$$[\bar{\xi}k\xi']' = k|\xi'|^2 + i\omega c|\xi|^2.$$

When we integrate with respect to x and use the boundary conditions we find that

$$0 = \int_A^B k|\xi'|^2 \, dx + i\omega \int_A^B c|\xi|^2 \, dx.$$

Thus

$$\int_A^B k|\xi'|^2 \, dx = 0$$

and, because $k(x)$ is positive, $\zeta'(x) = 0$. The boundary conditions now force $\zeta(x) = 0$.

Now let $\xi_A(x)$ and $\xi_B(x)$ be the solutions of the initial value problems

$$[k\xi'_A]' = i\omega c\xi_A, \qquad \xi_A(A) = 0, \qquad k(A)\xi'_A(A) = 1,$$

$$[k\xi'_B]' = i\omega c\xi_B, \qquad \xi_B(B) = 0, \qquad k(B)\xi'_B(B) = 1,$$

the existence and uniqueness of these solutions being guaranteed by a standard argument. In view of what was proved in the preceding paragraph, ξ_A and ξ_B are linearly independent on $[A, B]$, and, by a further standard argument,

$$\Gamma = k(\xi_A\xi'_B - \xi_B\xi'_A)$$

is a nonzero constant. Upon evaluating the right-hand side at each of the boundary points A and B, we see that

$$\Gamma = -\xi_B(A) = \xi_A(B).$$

"Variation of parameters" now tells us that

$$\Theta(x) = \frac{1}{\Gamma}(\theta_B\xi_A(x) - \theta_A\xi_B(x))$$

$$+ \frac{i\omega T_0}{\Gamma}\xi_A(x)\int_x^B \xi_B(y)\mu(y)\Phi'(y)\,dy$$

$$+ \frac{i\omega T_0}{\Gamma}\xi_B(x)\int_A^x \xi_A(y)\mu(y)\Phi'(y)\,dy,$$

or, upon introducing,

$$\eta_A(x) = \mu(x)\xi_A(x), \qquad \eta_B(x) = \mu(x)\xi_B(x),$$

that

$$\mu(x)\Theta(x) = \frac{1}{\Gamma}(\theta_B\eta_A(x) - \theta_A\eta_B(x))$$

$$+ \frac{i\omega T_0}{\Gamma}\eta_A(x)\int_x^B \eta_B(y)\Phi'(y)\,dy$$

$$+ \frac{i\omega T_0}{\Gamma}\eta_B(x)\int_A^x \eta_A(y)\Phi'(y)\,dy.$$

On using integration by parts and the boundary conditions $\eta_A(A) =$

$\eta_B(B) = 0$ we obtain the formula

$$\mu(x)\Theta(x) = \frac{1}{\Gamma}(\theta_B\eta_A(x) - \theta_A\eta_B(x))$$

$$- \frac{i\omega T_0}{\Gamma}\eta_A(x) \int_x^B \eta'_B(y)\Phi(y)\,dy$$

$$- \frac{i\omega T_0}{\Gamma}\eta_B(x) \int_A^x \eta'_A(y)\Phi(y)\,dy.$$

The next step is to calculate the derivative of the product $\mu\Theta$ and then to substitute into the differential equation

$$[\beta\Phi' - \mu\Theta]' = -\omega^2\rho\Phi,$$

thereby producing an integro-differential equation for Φ.

On noting that

$$\eta_A\eta'_B - \eta_B\eta'_A = \mu^2(\zeta_A\zeta'_B - \zeta_B\zeta'_A) = \frac{\Gamma\mu^2}{k},$$

we find the derivative to be

$$[\mu(x)\Theta(x)]' = \frac{1}{\Gamma}(\theta_B\eta'_A(x) - \theta_A\eta'_B(x))$$

$$+ i\omega T_0\frac{\mu(x)^2}{k(x)}\Phi(x) - \frac{i\omega T_0}{\Gamma}\int_A^B G(x, y)\Phi(y)\,dy,$$

the kernel being

$$G(x, y) = \begin{cases} \eta'_A(x)\eta'_B(y), & A \leq x \leq y \leq B, \\ \eta'_A(y)\eta'_B(x), & A \leq y \leq x \leq B. \end{cases}$$

Thus, we conclude that $\Phi(x)$ is a solution of the integro-differential equation

$$[\beta(x)\Phi'(x)]' = \frac{1}{\Gamma}(\theta_B\eta'_A(x) - \theta_A\eta'_B(x))$$

$$+ \left(-\omega^2\rho(x) + i\omega T_0\frac{\mu(x)^2}{k(x)}\right)\Phi(x)$$

$$- \frac{i\omega T_0}{\Gamma}\int_A^B G(x, y)\Phi(y)\,dy,$$

from which Θ is absent. This equation and the boundary conditions $\Phi(A) = \phi_A$, $\Phi(B) = \phi_B$ will be used to construct the required Fredholm integral equation.

As a preliminary step it is necessary to show that the boundary value problem

$$[\beta\zeta']' = \left(-\omega^2\rho + i\omega T_0\frac{\mu^2}{k}\right)\zeta, \qquad \zeta(A) = \zeta(B) = 0,$$

has no complex-valued solution other than $\zeta(x) = 0$.

It should be remarked that the presence of the term

$$i\omega T_0\frac{\mu^2}{k}$$

is essential if such a conclusion is to be valid; if the stress–temperature modulus $\mu(x)$ were to vanish identically, as it does within the isothermal linearized theory of elasticity, ζ would be the solution of a Sturm–Liouville problem for which the assertion is false whenever $\omega/2\pi$ is a characteristic frequency of vibration.

The proof is readily supplied. For, on multiplying the differential equation through by $\bar{\zeta}$, we find

$$[\bar{\zeta}\beta\zeta']' = \beta|\zeta'|^2 + \left(-\omega^2\rho + i\omega T_0\frac{\mu^2}{k}\right)|\zeta|^2,$$

and, on integrating with respect to x and using the fact that ζ vanishes at the boundary points, we obtain the equation

$$\int_A^B (\beta|\zeta'|^2 - \omega^2\rho|\zeta|^2)\,dx + i\omega T_0\int_A^B \frac{\mu^2}{k}|\zeta|^2\,dx = 0.$$

Hence

$$\int_A^B \frac{\mu^2}{k}|\zeta|^2\,dx = 0$$

if $\omega \neq 0$, and because $\mu(x)$ and $k(x)$ are positive, it must be that $\zeta(x) = 0$, as required. In the remaining case $\omega = 0$, the boundary value problem reduces to

$$[\beta\zeta']' = 0, \qquad \zeta(A) = \zeta(B) = 0,$$

and, because $\beta(x)$ is positive, we can conclude, just as in an earlier argument, that $\zeta(x) = 0$ in this case too.

Next, we choose $\zeta_A(x)$ and $\zeta_B(x)$ to be the solutions of the initial value problems

$$[\beta\zeta_A']' = \left(-\omega^2\rho + i\omega T_0\frac{\mu^2}{k}\right)\zeta_A, \qquad \zeta_A(A) = 0, \qquad \beta(A)\zeta_A'(A) = 1,$$

$$[\beta\zeta_B']' = \left(-\omega^2\rho + i\omega T_0\frac{\mu^2}{k}\right)\zeta_B, \qquad \zeta_B(B) = 0, \qquad \beta(B)\zeta_B'(B) = 1,$$

the existence and uniqueness of these solutions being guaranteed by standard arguments once again. What has just been proved ensures that $\zeta_A(x)$ and $\zeta_B(x)$ are linearly independent on $[A, B]$. Furthermore,

$$\Delta = \beta(\zeta_A\zeta_B' - \zeta_B\zeta_A')$$

is a nonzero constant, and

$$\Delta = -\zeta_B(A) = \zeta_A(B).$$

If, at this stage, we introduce the kernel

$$K(x, y) = \begin{cases} \zeta_A(x)\zeta_B(y), & A \leq x \leq y \leq B, \\ \zeta_A(y)\zeta_B(x), & A \leq y \leq x \leq B, \end{cases}$$

we can conclude from the integro-differential equation, and the boundary conditions $\Phi(A) = \phi_A$, $\Phi(B) = \phi_B$, that

$$\Phi(x) = \frac{1}{\Delta}(\phi_B\zeta_A(x) - \phi_A\zeta_B(x))$$

$$+ \frac{1}{\Gamma\Delta}\int_A^B K(x, y)(\theta_B\eta_A'(y) - \theta_A\eta_B'(y))\,dy$$

$$- \frac{i\omega T_0}{\Gamma\Delta}\int_A^B\int_A^B K(x, y)G(y, z)\Phi(z)\,dz\,dy.$$

In brief, $\Phi(x)$ must satisfy the Fredholm integral equation

$$\Phi(x) = \frac{1}{\Delta}(\phi_B\zeta_A(x) - \phi_A\zeta_B(x))$$

$$+ \frac{1}{\Gamma\Delta}\int_A^B K(x, y)(\theta_B\eta_A'(y) - \theta_A\eta_B'(y))\,dy$$

$$- \frac{i\omega T_0}{\Gamma\Delta}\int_A^B L(x, y)\Phi(y)\,dy,$$

whose kernel is

$$L(x, y) = \int_A^B K(x, z)G(z, y) \, dz.$$

The steps in the argument are reversible and, indeed, if $\Phi(x)$ is a solution of the integral equation, and $\Theta(x)$ is defined by

$$\mu(x)\Theta(x) = \frac{1}{\Gamma}(\theta_B \eta_A(x) - \theta_A \eta_B(x))$$

$$- \frac{i\omega T_0}{\Gamma}\eta_A(x) \int_x^B \eta_B'(y)\Phi(y) \, dy$$

$$- \frac{i\omega T_0}{\Gamma}\eta_B(x) \int_A^x \eta_A'(y)\Phi(y) \, dy,$$

then $\Phi(x)$ and $\Theta(x)$ satisfy the differential equations and the boundary conditions of §36.

Fredholm's theory tells us that the integral equation has a unique solution if and only if the homogeneous equation

$$\Phi(x) = -\frac{i\omega T_0}{\Gamma\Delta} \int_A^B L(x, y)\Phi(y) \, dy$$

has no solution other than $\Phi(x) = 0$. However, the homogeneous equation is what is obtained by setting

$$\phi_A = \phi_B = \theta_A = \theta_B = 0,$$

and in that case our proof of uniqueness establishes that $\Phi(x) = 0$. Thus, $\Phi(x)$ and $\Theta(x)$ exist and are uniquely defined, as §36 asserts, whatever the values of ϕ_A, ϕ_B, θ_A, θ_B.

39. In order to achieve the main objective of this chapter, namely to establish the correctness of what §35 asserts, we shall be forced to examine more closely the behaviour of Φ and Θ when $\omega \to 0$ and the boundary values ϕ_A, ϕ_B, θ_A, θ_B are held fixed. It is convenient to use the notation

$$\Phi(x, \omega), \quad \Theta(x, \omega),$$

thereby emphasizing the dependence upon ω but suppressing the dependence upon the boundary values.

What we shall need to check is that:

When $\omega \to 0$, the order relations

$$\Phi(x, \omega) = \Phi(x, 0) + O(|\omega|),$$

$$\Theta(x, \omega) = \Theta(x, 0) + O(|\omega|),$$

$$\Phi'(x, \omega) = \Phi'(x, 0) + O(|\omega|),$$

$$\Theta'(x, \omega) = \Theta'(x, 0) + O(|\omega|)$$

are valid uniformly with respect to x in $[A, B]$.

If we set $\omega = 0$ in the differential equations of §36 we see that $\Phi(x, 0)$ and $\Theta(x, 0)$ are the solutions of the boundary value problem

$$[\beta(x)\Phi'(x, 0) - \mu(x)\Theta(x, 0)]' = 0,$$

$$[k(x)\Theta'(x, 0)]' = 0,$$

$$\Phi(A, 0) = \phi_A, \qquad \Phi(B, 0) = \phi_B, \qquad \Theta(A, 0) = \theta_A, \qquad \Theta(B, 0) = \theta_B,$$

and, hence, their values can be found from the formulae

$$\int_A^B \frac{dy}{\beta(y)} \cdot \Phi(x, 0) = \phi_A \int_x^B \frac{dy}{\beta(y)} + \phi_A \int_A^x \frac{dy}{\beta(y)}$$

$$- \int_A^x \frac{dy}{\beta(y)} \cdot \int_x^B \frac{\mu(y)}{\beta(y)} \Theta(y, 0) \, dy$$

$$+ \int_x^B \frac{dy}{\beta(y)} \cdot \int_A^x \frac{\mu(y)}{\beta(y)} \Theta(y, 0) \, dy,$$

$$\int_A^B \frac{dy}{\beta(y)} \cdot \Theta(x, 0) = \theta_A \int_x^B \frac{dy}{k(y)} + \theta_B \int_A^x \frac{dy}{k(y)}.$$

It follows from the second of these that $\Theta'(x, 0) = 0$ if $\theta_A = \theta_B$ and so we have the following important consequence of the last of the four order relations.

If $\theta_A = \theta_B$ then, as $\omega \to 0$,

$$\Theta'(x, \omega) = O(|\omega|),$$

uniformly with respect to x in $[A, B]$.

The procedure for deriving the order relations is straightforward in principle but somewhat tedious in detail. It involves replacing the boundary value problem of §36 by a pair of coupled Fredholm integral equations; the method is different from that used in §38 to establish existence, for the kernels $G(x, y)$, $K(x, y)$, $L(x, y)$ which appear there depend upon ω and do so in very complicated ways.

To begin, we transfer the term $[\mu\Theta]'$ to the right-hand side of the first of the differential equations of §36, and construct a Green's function associated with the operator

$$\frac{d}{dx}\left[\beta(x)\frac{d}{dx}\right],$$

to obtain the expression

$$\int_A^B \frac{dy}{\beta(y)} \cdot \Phi(x, \omega)$$

$$= \phi_A \int_x^B \frac{dy}{\beta(y)} + \phi_B \int_A^x \frac{dy}{\beta(y)}$$

$$- \int_A^x \frac{dy}{\beta(y)} \cdot \int_x^B \int_y^B \frac{([\mu(y)\Theta(y, \omega)]' - \omega^2 \rho(y)\Phi(y, \omega))}{\beta(z)} \, dz \, dy$$

$$- \int_x^B \frac{dy}{\beta(y)} \cdot \int_A^x \int_A^y \frac{([\mu(y)\Theta(y, \omega)]' - \omega^2 \rho(y)\Phi(y, \omega))}{\beta(z)} \, dz \, dy.$$

Next, we use integration by parts in the integrals which contain the derivative $[\mu\Theta]'$ and arrive at the expression

$$\int_A^B \frac{dy}{\beta(y)} \cdot \Phi(x, \omega)$$

$$= \phi_A \int_x^B \frac{dy}{\beta(y)} + \phi_B \int_A^x \frac{dy}{\beta(y)}$$

$$- \int_A^x \frac{dy}{\beta(y)} \cdot \int_x^B \frac{\mu(y)}{\beta(y)} \Theta(y, \omega) \, dy$$

$$+ \int_x^B \frac{dy}{\beta(y)} \cdot \int_A^x \frac{\mu(y)}{\beta(y)} \Theta(y, \omega) \, dy$$

$$+ \omega^2 \int_A^x \frac{dy}{\beta(y)} \cdot \int_x^B \int_y^B \frac{\rho(y)}{\beta(z)} \Phi(y, \omega) \, dz \, dy$$

$$+ \omega^2 \int_x^B \frac{dy}{\beta(y)} \cdot \int_A^x \int_A^y \frac{\rho(y)}{\beta(z)} \Phi(y, \omega) \, dz \, dy,$$

from which the derivatives of Φ and Θ are absent. Thus, if we make use of the formulae for $\Phi(x, 0)$ and $\Theta(x, 0)$ that have been derived already we deduce the first of our pair of Fredholm equations, namely

$$\int_A^B \frac{dy}{\beta(y)} \cdot (\Phi(x, \omega) - \Phi(x, 0))$$

$$= -\int_A^x \frac{dy}{\beta(y)} \cdot \int_x^B \frac{\mu(y)}{\beta(y)} (\Theta(y, \omega) - \Theta(y, 0)) \, dy$$

$$+ \int_x^B \frac{dy}{\beta(y)} \cdot \int_A^x \frac{\mu(y)}{\beta(y)} (\Theta(y, \omega) - \Theta(y, 0)) \, dy$$

$$+ \omega^2 \int_A^x \frac{dy}{\beta(y)} \cdot \int_x^B \int_y^B \frac{\rho(y)}{\beta(z)} \Phi(y, \omega) \, dz \, dy$$

$$+ \omega^2 \int_x^B \frac{dy}{\beta(y)} \cdot \int_A^x \int_A^y \frac{\rho(y)}{\beta(z)} \Phi(y, \omega) \, dz \, dy.$$

In much the same way, if we turn to the second of the differential equations of §36, and construct a Green's function associated with the operator

$$\frac{d}{dx}\left[k(x) \frac{d}{dx} \right],$$

we see that

$$\int_A^B \frac{dy}{k(y)} \cdot \Theta(x, \omega)$$

$$= \theta_A \int_x^B \frac{dy}{k(y)} + \theta_B \int_A^x \frac{dy}{k(y)}$$

$$- i\omega \int_A^x \frac{dy}{k(y)} \cdot \int_x^B \int_y^B \frac{(T_0 \mu(y)\Phi'(y, \omega) + c(y)\Theta(y, \omega))}{k(z)} \, dz \, dy$$

$$- i\omega \int_x^B \frac{dy}{k(y)} \cdot \int_A^x \int_A^y \frac{(T_0 \mu(y)\Phi'(y, \omega) + c(y)\Theta(y, \omega))}{k(z)} \, dz \, dy.$$

An integration by parts, and use of the formula for $\Theta(x, 0)$, now yield the second of our pair of Fredholm equations, namely

$$\int_A^B \frac{dy}{k(y)} \cdot (\Theta(x, \omega) - \Theta(x, 0))$$

$$= i\omega T_0 \int_A^x \frac{dy}{k(y)} \cdot \int_x^B \left[\mu'(y) \int_y^B \frac{dz}{k(z)} - \frac{\mu(y)}{k(y)} \right] \Phi(y, \omega) \, dy$$

$$- i\omega T_0 \int_x^B \frac{dy}{k(y)} \cdot \int_A^x \left[\mu'(y) \int_A^y \frac{dz}{k(z)} + \frac{\mu(y)}{k(y)} \right] \Phi(y, \omega)\, dy$$

$$- i\omega \int_A^x \frac{dy}{k(y)} \cdot \int_x^B \int_y^B \frac{c(y)}{k(z)} \Theta(y, \omega)\, dz\, dy$$

$$- i\omega \int_x^B \frac{dy}{k(y)} \cdot \int_A^x \int_A^y \frac{c(y)}{k(z)} \Theta(y, \omega)\, dz\, dy.$$

Once again, the derivatives of Φ and Θ are absent from the equation.

In order to complete the proof we introduce the norms:

$$\|\Phi_\omega\| = \text{Max}\{|\Phi(x, \omega)|: A \leq x \leq B\},$$

$$\|\Theta_\omega\| = \text{Max}\{|\Theta(x, \omega)|: A \leq x \leq B\},$$

$$\|\Phi_\omega - \Phi_0\| = \text{Max}\{|\Phi(x, \omega) - \Phi(x, 0)|: A \leq x \leq B\},$$

$$\|\Theta_\omega - \Theta_0\| = \text{Max}\{|\Theta(x, \omega) - \Theta(x, 0)|: A \leq x \leq B\}.$$

Upon estimating the right-hand sides of the Fredholm equations we obtain the inequalities

$$\|\Phi_\omega - \Phi_0\| \leq \kappa \|\Theta_\omega - \Theta_0\| + \omega^2 \kappa \|\Phi_\omega\|,$$

$$\|\Theta_\omega - \Theta_0\| \leq |\omega| \kappa (\|\Phi_\omega\| + \|\Theta_\omega\|),$$

κ being a sufficiently large positive constant which is independent of ω.

The first can be replaced by the weaker inequality

$$\|\Phi_\omega - \Phi_0\| \leq \kappa \|\Theta_\omega - \Theta_0\| + \omega^2 \kappa (\|\Phi_\omega\| + \|\Theta_\omega\|)$$

and, with the aid of the second, this in turn implies

$$\|\Phi_\omega - \Phi_0\| \leq (|\omega| \kappa^2 + \omega^2 \kappa)(\|\Phi_\omega\| + \|\Theta_\omega\|).$$

Thus, if we invoke the triangle inequalities

$$\|\Phi_\omega\| \leq \|\Phi_0\| + \|\Phi_\omega - \Phi_0\|, \qquad \|\Theta_\omega\| \leq \|\Theta_0\| + \|\Theta_\omega - \Theta_0\|,$$

we find

$$\|\Phi_\omega - \Phi_0\| \leq (|\omega| \kappa^2 + \omega^2 \kappa)(\|\Phi_0\| + \|\Theta_0\| + \|\Phi_\omega - \Phi_0\|$$
$$+ \|\Theta_\omega - \Theta_0\|),$$

$$\|\Theta_\omega - \Theta_0\| \leq |\omega| \kappa (\|\Phi_0\| + \|\Theta_0\| + \|\Phi_\omega - \Phi_0\| + \|\Theta_\omega - \Theta_0\|).$$

Adding the last two inequalities tells us that

$$\|\Phi_\omega - \Phi_0\| + \|\Theta_\omega - \Theta_0\| \le \frac{(|\omega|(\kappa + \kappa^2) + \omega^2\kappa)}{1 - |\omega|(\kappa + \kappa^2) - \omega^2\kappa}(\|\Phi_0\| + \|\Theta_0\|)$$

provided ω is so small that $|\omega|(\kappa + \kappa^2) + \omega^2\kappa < 1$. Hence

$$\|\Phi_\omega - \Phi_0\| = O(|\omega|), \qquad \|\Theta_\omega - \Theta_0\| = O(|\omega|),$$

or, in other words, the order relations

$$\Phi(x, \omega) = \Phi(x, 0) + O(|\omega|),$$

$$\Theta(x, \omega) = \Theta(x, 0) + O(|\omega|),$$

are valid uniformly on $[A, B]$.

Finally, we return to the first of the Fredholm equations of this section, differentiate throughout with respect to x, and obtain the formula

$$\int_A^B \frac{dy}{\beta(y)} \cdot (\Phi'(x, \omega) - \Phi'(x, 0))$$

$$= -\frac{1}{\beta(x)} \int_A^B \frac{\mu(y)}{\beta(y)} (\Theta(y, \omega) - \Theta(y, 0)) \, dy$$

$$+ \left(\int_A^B \frac{dy}{\beta(y)} \right) \frac{\mu(x)}{\beta(x)} (\Theta(x, \omega) - \Theta(x, 0))$$

$$+ \frac{\omega^2}{\beta(x)} \left(\int_x^B \int_y^B - \int_A^x \int_A^y \right) \frac{\rho(y)}{\beta(z)} \Phi(y, \omega) \, dz \, dy.$$

By virtue of what has just been proved, the right-hand side is uniformly $O(|\omega|)$ and so

$$\Phi'(x, \omega) = \Phi'(x, 0) + O(|\omega|)$$

uniformly in $[A, B]$.

In much the same way, if we differentiate the second Fredholm equation with respect to x we can deduce that

$$\Theta'(x, \omega) = \Theta'(x, 0) + O(|\omega|)$$

uniformly in $[A, B]$, and thus the proof of the order relations is complete.

40. As the next step in the proof of the main result of this chapter, namely what §35 asserts to be true, we turn to considering periodic solutions of the linearized displacement–temperature equations which can be maintained by supplying periodic displacements $u(A, t)$ and $u(B, t)$ at the boundary points, and a periodic temperature $\tau(t)$ $(= T(A, t) = T(B, t))$ which is common to both boundary points.

Thus, let $P_A(s), P_B(s), P(s)$ be any real trigonometric polynomials of the forms

$$P_A(s) = \text{Re} \sum A_n \exp(ins),$$

$$P_B(s) = \text{Re} \sum B_n \exp(ins),$$

$$P(s) = \text{Re} \sum C_n \exp(ins),$$

where the coefficients A_n, B_n, C_n may be any complex numbers, and each sum is taken over a finite set of integers n. By defining appropriately many of A_n, B_n, C_n to be zero, it can be arranged that each sum is taken over the same finite set, and, henceforth, this will be supposed to be the case. Each of the polynomials is periodic in s, with period 2π.

It will be necessary to reintroduce the expanded notation

$$\Phi(x, \omega, \phi_A, \phi_B, \theta_A, \theta_B),$$

$$\Theta(x, \omega, \phi_A, \phi_B, \theta_A, \theta_B)$$

of §36; in what follows θ_A and θ_B always coincide.

If α is any number in $0 < \alpha < 1$, and if we set

$$u(x, t) = \text{Re} \sum \Phi(x, n\alpha, A_n, B_n, C_n, C_n) \exp(in\alpha t),$$

$$T(x, t) = T_0 + \text{Re} \sum \Theta(x, n\alpha, A_n, B_n, C_n, C_n) \exp(in\alpha t),$$

we have constructed a displacement field and a temperature field which are C^2 and satisfy the linearized displacement–temperature equations (with $f = h = 0$), which are periodic with period $2\pi/\alpha$ in their dependence upon t, that is to say

$$u\left(x, t + \frac{2\pi}{\alpha}\right) = u(x, t), \qquad T\left(x, t + \frac{2\pi}{\alpha}\right) = T(x, t),$$

and which satisfy the boundary conditions

$$u(A, t) = P_A(\alpha t), \qquad u(B, t) = P_B(\alpha t),$$

$$T(A, t) = T(B, t) = T_0 + P(\alpha t).$$

The trigonometric polynomials

$$P_A(\alpha t), \quad P_B(\alpha t), \quad P(\alpha t)$$

are obtained from

$$P_A(s), \quad P_B(s), \quad P(s)$$

by retardation; that is to say, as t and s vary the members of the first trio run through the same sets of values as do their counterparts in the second trio, but they do so at slower rates.

It is proposed to examine the behaviour, in the limit of extreme retardation, of the work done by the body on the interval $[0, 2\pi/\alpha]$, the heat absorbed by the body on that interval, and the efficiency.

The statement of what happens involves the numbers

$$\aleph = \int_A^B \frac{\mu}{\beta}\, dx \bigg/ \int_A^B \frac{1}{\beta}\, dx,$$

$$\beth = \left[\int_A^B \frac{\mu^2}{\beta}\, dx \cdot \int_A^B \frac{1}{\beta}\, dx - \left(\int_A^B \frac{\mu}{\beta}\, dx \right)^2 + \frac{1}{T_0} \int_A^B c\, dx \right] \bigg/ \int_A^B \frac{1}{\beta}\, dx,$$

each of which is positive, the latter being so by virtue of the Schwarz inequality.

In the limit of extreme retardation, that is when $\alpha \to 0$,

$$\int_0^{2\pi/\alpha} W(t)\, dt \to \aleph \int_0^{2\pi} P(s) \frac{d}{ds}(P_B(s) - P_A(s))\, ds,$$

$$\int_0^{2\pi/\alpha} Q^+(t)\, dt \to \tfrac{1}{2} T_0 \int_0^{2\pi} \left| \frac{d}{ds}(\aleph(P_B(s) - P_A(s)) + \beth P(s)) \right| ds.$$

41. The proof of what happens under extreme retardation will be made to depend upon two identities of §32 and a further identity of §33.

It should be noted that hypotheses (i), (ii), and (iii) of §32 are satisfied here. Indeed, (i) holds because $u(x, t)$ and $T(x, t)$ satisfy the linearized displacement–temperature equations, with $f = h = 0$. Again (ii) holds, with $\tau(t) = T_0 + P(\alpha t)$, and (iii) holds, with $t_1 = 0$ and $t_2 = 2\pi/\alpha$, because of periodicity. Hence,

$$\int_0^{2\pi/\alpha} Q(t)\, dt = 0,$$

$$\int_0^{2\pi/\alpha} W(t)\, dt + \frac{1}{T_0} \int_0^{2\pi/\alpha} \int_A^B g(x, t)q(x, t)\, dx\, dt$$

$$= \frac{1}{T_0} \int_0^{2\pi/\alpha} (T_0 + P(\alpha t))Q(t)\, dt,$$

the net heat flux into the body being

$$Q(t) = T_0 \int_A^B \overline{\dot{S}(x, t)} \, dx.$$

To derive the limiting value of the heat absorbed by the body we start from the observation that

$$Q^+ = \text{Max}(Q, 0) = \tfrac{1}{2}(|Q| + Q).$$

As the net heat gained by the body on $[0, 2\pi/\alpha]$ vanishes, it must be that

$$\int_0^{2\pi/\alpha} Q^+(t) \, dt = \tfrac{1}{2} \int_0^{2\pi/\alpha} |Q(t)| \, dt$$

and, hence,

$$\int_0^{2\pi/\alpha} Q^+(t) \, dt = \tfrac{1}{2} T_0 \int_0^{2\pi/\alpha} \left| \overline{\int_A^B \dot{S}(x, t) \, dx} \right| dt.$$

The integral

$$
\begin{aligned}
\int_A^B S \, dx &= \int_A^B \left(S_0 + \mu E + \frac{c}{T_0}(T - T_0) \right) dx \\
&= \int_A^B S_0(x) \, dx + \text{Re} \left\{ \sum \int_A^B \left[\mu(x)\Phi'(x, n\alpha, A_n, B_n, C_n, C_n) \right.\right. \\
&\qquad \left.\left. + \frac{c(x)}{T_0} \Theta(x, n\alpha, A_n, B_n, C_n, C_n) \right] dx \cdot \exp(in\alpha t) \right\}.
\end{aligned}
$$

Thus,

$$
\begin{aligned}
T_0 \int_A^B \dot{S} \, dx = \text{Re} \Big\{ \sum \int_A^B & [T_0\mu(x)\Phi'(x, n\alpha, A_n, B_n, C_n, C_n) \\
& + c(x)\Theta(x, n\alpha, A_n, B_n, C_n, C_n)] \, dx \cdot in\alpha \exp(in\alpha t) \Big\}
\end{aligned}
$$

and the heat absorbed by the body is

$$
\begin{aligned}
&\int_0^{2\pi/\alpha} Q^+(t) \, dt \\
&= \tfrac{1}{2} \int_0^{2\pi/\alpha} \Big| \text{Re} \Big\{ \sum \int_A^B [T_0\mu(x)\Phi'(x, n\alpha, A_n, B_n, C_n, C_n) \\
&\qquad + c(x)\Theta(x, n\alpha, A_n, B_n, C_n, C_n)] \, dx \cdot in\alpha \exp(in\alpha t) \Big\} \Big| \, dt
\end{aligned}
$$

$$= \frac{1}{2} \int_0^{2\pi} \left| \text{Re} \left\{ \sum \int_A^B [T_0 \mu(x) \Phi'(x, n\alpha, A_n, B_n, C_n, C_n) \right. \right.$$

$$\left. \left. + c(x) \Theta(x, n\alpha, A_n, B_n, C_n, C_n)] \, dx \cdot in \, \exp(ins) \right\} \right| ds,$$

where we have made the change of variable $s = \alpha t$.

In view of the order relations of §39, and because the set of integers over which the summation extends is finite, it must be that, when $\alpha \to 0$, the heat absorbed by the body converges to

$$\frac{1}{2} \int_0^{2\pi} \left| \text{Re} \left\{ \sum \int_A^B [T_0 \mu(x) \Phi'(x, 0, A_n, B_n, C_n, C_n) \right. \right.$$

$$\left. \left. + c(x) \Theta(x, 0, A_n, B_n, C_n, C_n)] \, dx \cdot in \, \exp(ins) \right\} \right| ds.$$

Since, as §39 tells us,

$$\int_A^B \frac{dy}{\beta(y)} \cdot \Phi'(x, 0, A_n, B_n, C_n, C_n)$$

$$= \frac{B_n - A_n}{\beta(x)} + C_n \left[\frac{\mu(x)}{\beta(x)} \int_A^B \frac{dy}{\beta(y)} - \frac{1}{\beta(x)} \int_A^B \frac{\mu(y)}{\beta(y)} \, dy \right],$$

$$\Theta(x, 0, A_n, B_n, C_n, C_n) = C_n,$$

the integral

$$\int_A^B [T_0 \mu(x) \Phi'(x, 0, A_n, B_n, C_n, C_n) + c(x) \Theta(x, 0, A_n, B_n, C_n, C_n)] \, dx$$

$$= \left[T_0 \int_A^B \frac{\mu}{\beta} \, dx (B_n - A_n) + \left(T_0 \int_A^B \frac{\mu^2}{\beta} \, dx \int_A^B \frac{1}{\beta} \, dx \right. \right.$$

$$\left. \left. - T_0 \left(\int_A^B \frac{\mu}{\beta} \, dx \right)^2 + \int_A^B c \, dx \right) C_n \right] \bigg/ \int_A^B \frac{1}{\beta} \, dx$$

$$= T_0 \aleph (B_n - A_n) + T_0 \beth C_n$$

and, therefore,

$$\int_0^{2\pi/\alpha} Q^+(t) \, dt \to \frac{1}{2} T_0 \int_0^{2\pi} \left| \text{Re} \sum (\aleph(B_n - A_n) + \beth C_n) \, in \, \exp(ins) \right| ds$$

$$= \frac{1}{2} T_0 \int_0^{2\pi} \left| \frac{d}{ds} (\aleph(P_B(s) - P_A(s)) + \beth P(s)) \right| ds.$$

In order to ascertain the limiting behaviour of the work done by the body, we return to the identity

$$\int_0^{2\pi/\alpha} W(t)\, dt + \frac{1}{T_0}\int_0^{2\pi/\alpha}\int_A^B g(x, t)q(x, t)\, dx\, dt$$

$$= \frac{1}{T_0}\int_0^{2\pi/\alpha}(T_0 + P(\alpha t))Q(t)\, dt$$

and use the same sort of argument as that just deployed.

Since the net heat gained by the body is known to vanish, the right-hand side coincides with

$$\frac{1}{T_0}\int_0^{2\pi/\alpha} P(\alpha t)Q(t)\, dt$$

and this in turn equals

$$\int_0^{2\pi/\alpha} P(\alpha t)\overline{\left(\int_A^B \dot{S}(x, t)\, dx\right)} dt$$

$$= \int_0^{2\pi/\alpha} P(\alpha t)\,\mathrm{Re}\left\{\sum \int_A^B \left[\mu(x)\Phi'(x, n\alpha, A_n, B_n, C_n, C_n)\right.\right.$$

$$\left.\left. + \frac{c(x)}{T_0}\Theta(x, n\alpha, A_n, B_n, C_n, C_n)\right] dx \cdot in\alpha \exp(in\alpha t)\right\} dt$$

$$= \int_0^{2\pi} P(s)\,\mathrm{Re}\left\{\sum \int_A^B \left[\mu(x)\Phi'(x, n\alpha, A_n, B_n, C_n, C_n)\right.\right.$$

$$\left.\left. + \frac{c(x)}{T_0}\Theta(x, n\alpha, A_n, B_n, C_n, C_n)\right] dx \cdot in \exp(ins)\right\} ds.$$

When $\alpha \to 0$, this last expression converges to

$$\int_0^{2\pi} P(s)\frac{d}{ds}(\aleph(P_B(s) - P_A(s)) + \beth P(s))\, ds,$$

which, because

$$\int_0^{2\pi} P(s)\frac{d}{ds}P(s)\, ds = \tfrac{1}{2}[P(s)^2]_0^{2\pi} = 0,$$

reduces to

$$\aleph \int_0^{2\pi} P(s)\frac{d}{ds}(P_B(s) - P_A(s))\, ds.$$

On the other hand, the double integral

$$\int_0^{2\pi/\alpha} \int_A^B g(x, t)q(x, t) \, dx \, dt$$

$$= \int_0^{2\pi/\alpha} \int_A^B k(x)g(x, t)^2 \, dx \, dt$$

$$= \int_0^{2\pi/\alpha} \int_A^B k(x) \left[\operatorname{Re} \sum \Theta'(x, n\alpha, A_n, B_n, C_n, C_n) \exp(in\alpha t) \right]^2 \, dx \, dt$$

$$= \frac{1}{\alpha} \int_0^{2\pi} \int_A^B k(x) \left[\operatorname{Re} \sum \Theta'(x, n\alpha, A_n, B_n, C_n, C_n) \exp(ins) \right]^2 \, dx \, ds$$

$$\leqq \frac{1}{\alpha} \int_0^{2\pi} \int_A^B k(x)|\sum \Theta'(x, n\alpha, A_n, B_n, C_n, C_n) \exp(ins)|^2 \, dx \, ds$$

$$\leqq \frac{1}{\alpha} \int_0^{2\pi} \int_A^B k(x)(\sum |\Theta'(x, n\alpha, A_n, B_n, C_n, C_n)|)^2 \, dx \, ds$$

$$= \frac{2\pi}{\alpha} \int_A^B k(x)(\sum |\Theta'(x, n\alpha, A_n, B_n, C_n, C_n)|)^2 \, dx.$$

In view of an order relation of §39, and the fact that only finitely many terms occur in the sum,

$$\int_A^B k(x)(\sum |\Theta'(x, n\alpha, A_n, B_n, C_n, C_n)|)^2 \, dx = O(\alpha^2).$$

Thus the double integral is $O(\alpha)$, and, therefore, tends to zero when $\alpha \to 0$. We conclude finally that, when $\alpha \to 0$, the work done by the body converges to the limit that was advertised, namely

$$\aleph \int_0^{2\pi} P(s)\frac{d}{ds}(P_B(s) - P_A(s)) \, ds.$$

42. We are in position, at last, to complete the proof of the result stated in §35 to the effect that the bound

$$\frac{M - m}{T_0}$$

cannot be replaced by any smaller number. The argument involves defining the displacement field $u(x, t)$ and the temperature field $T(x, t)$ as in §40, and making special choices of the trigonometric polynomials $P_A(s)$, $P_B(s)$, $P(s)$.

It will do, in fact, to choose

$$P_A(s) = 0,$$

$$P_B(s) = -\frac{\daleth}{2\aleph}(M + m + (M - m)\cos(s))$$

$$+ \frac{\delta}{2^N\aleph}\sum_{n=0}^{N}\frac{1}{N - 2n}\binom{N}{n}\sin((N - 2n)s),$$

$$P(s) = -T_0 + \tfrac{1}{2}(M + m) + \tfrac{1}{2}(M - m)\cos(s),$$

where N is a sufficiently large *odd* positive integer, and δ is any positive number.

The interval $[t_1, t_2]$ is chosen to be $[0, 2\pi/\alpha]$, where α is a sufficiently small positive number.

These choices ensure that requirement (i) of §35 is satisfied. Requirement (ii) is satisfied, with

$$\tau(t) = T_0 + P(\alpha t) = \tfrac{1}{2}(M + m) + \tfrac{1}{2}(M - m)\cos(\alpha t),$$

and, therefore, (iv) is satisfied as well. Moreover, (iii) is satisfied by virtue of the periodicity of u and T. Thus it remains to check (v) and (vi).

Because of the way $P_A(s)$, $P_B(s)$, $P(s)$ have been chosen,

$$\frac{d}{ds}(\aleph(P_B(s) - P_A(s)) + \daleth P(s))$$

$$= \frac{\delta}{2^N}\sum_{n=0}^{N}\binom{N}{n}\cos((N - 2n)s) = \delta\cos^N(s)$$

and, in terms of the notation of §30, which sets

$$I_n = \int_0^{\pi/2}\cos^n(s)\,ds,$$

we have

$$\int_0^{2\pi}\left|\frac{d}{ds}(\aleph(P_B(s) - P_A(s)) + \daleth P(s))\right|ds = \delta\int_0^{2\pi}|\cos^N(s)|\,ds = 4\delta I_N.$$

As $P(s)$ is periodic, the integral

$$\aleph \int_0^{2\pi} P(s)\frac{d}{ds}(P_B(s) - P_A(s))\, ds$$

$$\doteq \int_0^{2\pi} P(s)\frac{d}{ds}(\aleph(P_B(s) - P_A(s)) + \beth P(s))\, ds$$

$$= \int_0^{2\pi} \delta \cos^N(s)(-T_0 + \tfrac{1}{2}(M + m) + \tfrac{1}{2}(M - m)\cos(s))\, ds$$

and, because N is odd, this equals

$$\tfrac{1}{2}\delta(M - m) \int_0^{2\pi} \cos^{N+1}(s)\, ds = 2\delta(M - m)I_{N+1}.$$

It now follows with the aid of the results of §§40 and 41 that, when $\alpha \to 0$,

$$\int_0^{2\pi/\alpha} Q^+(t)\, dt \to 2T_0\delta I_N > 0,$$

$$\frac{\int_0^{2\pi/\alpha} W(t)\, dt}{\int_0^{2\pi/\alpha} Q^+(t)\, dt} \to \frac{2\delta(M - m)I_{N+1}}{2T_0\delta I_N} = \frac{(M - m)}{T_0}\cdot\frac{I_{N+1}}{I_N}.$$

Let ε be the arbitrary positive number that is supposed given in §35, and let the odd positive integer N be chosen sufficiently large as to ensure that

$$1 - \frac{\varepsilon T_0}{2(M - m)} < \frac{I_{N+1}}{I_N} < 1.$$

Then

$$\lim_{\alpha \to 0} \int_0^{2\pi/\alpha} Q^+(t)\, dt > 0,$$

$$\lim_{\alpha \to 0} \frac{\int_0^{2\pi/\alpha} W(t)\, dt}{\int_0^{2\pi/\alpha} Q^+(t)\, dt} > \frac{M - m}{T_0} - \tfrac{1}{2}\varepsilon,$$

and, hence, by choosing α sufficiently small it can be arranged that both the inequalities

$$\int_0^{2\pi/\alpha} Q^+(t)\, dt > 0,$$

$$\int_0^{2\pi/\alpha} W(t)\, dt > \left(\frac{M - m}{T_0} - \varepsilon\right)\int_0^{2\pi/\alpha} Q^+(t)\, dt$$

are satisfied simultaneously. Thus, requirements (v) and (vi) of §35 are satisfied, and the proof of what §35 asserts is complete.

43. It should be observed that the line of argument in this chapter is effective for the reason that the choices of boundary displacements and of boundary temperature ensure that $Q(t)$, the net heat flux into the body, is approximately equal to

$$\delta\alpha \cos^N(\alpha t).$$

The values of t at which this function attains its maximum and minimum values coincide with those at which the boundary temperature

$$\tau(t) = \tfrac{1}{2}(M + m) + \tfrac{1}{2}(M - m)\cos(\alpha t)$$

attains its maximum and minimum values. When N is large and odd, the bulk of the graph of $Q(t)$ is concentrated into narrow peaks of height $+\delta\alpha$ and narrow troughs of height $-\delta\alpha$. Thus, when the body absorbs heat, that is when $Q(t) > 0$, the boundary temperature is close to its maximum value M, and when the body emits heat, that is when $Q(t) < 0$, the boundary temperature is close to its minimum value m. In other words, by controlling just the boundary displacements and the boundary temperature, it has been arranged that the body shall execute what is approximately a quasi-static Carnot cycle.

CHAPTER 6

Versions of a Second Law of Thermodynamics

44. The line of argument adopted in this tract has avoided commitment to any particular statement of a second law of thermodynamics, or even commitment as to whether there is a second law that is of wide or universal applicability. The price to be paid for this approach is the necessity of having to postulate at the outset that the equations of state

$$\hat{\sigma} = \partial_E \hat{F},$$

$$\hat{U} = \hat{F} - T\partial_T \hat{F},$$

$$\hat{S} = -\partial_T \hat{F}$$

are valid, as is the heat conduction inequality

$$gq > 0 \qquad \text{if} \quad g \neq 0.$$

It might well be thought desirable to deduce the equations of state and the heat conduction inequality from more primitive postulates, but the fact remains that I have not taken such a course.

I intend instead to reverse the line of argument and examine to what extent certain statements of a second law, which were proposed by the pioneers of thermodynamics, are implied by the theories considered here. Three examples of such statements are listed below.

It is impossible for a self-acting machine, unaided by any external agency, to convey heat from one body to another at a higher temperature. (Clausius–Kelvin, as stated by Kelvin [7].)

It is impossible to construct a machine which functions with a regular period and which does nothing but raise a weight and cause a corresponding cooling of a heat reservoir. (Planck [11].)

An isolated system, in which thermal effects are present, cannot return to a former state. (Perrin [10].)

One of the difficulties which dogs any attempt to interpret these versions of a second law is that they are all couched in purely verbal terms, and somewhat obscure ones at that. Thus, it cannot be claimed that a mathematical restatement of, let us say, Planck's version necessarily reflects exactly what Planck had in mind; the most that is claimed here is that our mathematical restatement is a logical consequence of nonlinear thermoelasticity, or of homogeneous and dissipationless thermoelasticity, or of linearized thermoelasticity, and operates within a framework of ideas which is similar to that of Planck's.

Part 1. Nonlinear Thermoelasticity

45. The theory of Chapter 1 is required to be in force throughout §§45–48. The first version of a second law to be established is a rather sophisticated one.

The Clausius–Duhem inequality

$$\overline{\int_A^B S \, dx} \geqq \left[\frac{q}{T}\right]_A^B + \int_A^B \frac{h}{T} \, dx$$

is valid.

In fact, more is true than has been claimed for the inequality

$$\overline{\int_a^b S \, dx} \geqq \left[\frac{q}{T}\right]_a^b + \int_a^b \frac{h}{T} \, dx$$

is valid for every subbody $[a, b]$. The counterpart to this assertion for three-dimensional bodies was taken as the cornerstone for the building of rational continuum thermodynamics by Coleman and Noll [3]; many authors subsequently adopted the same starting point.

The Clausius–Duhem inequality is a straightforward consequence of a result established in §9, namely the equation

$$\dot{S} = \left(\frac{q}{T}\right)' + \frac{gq}{T^2} + \frac{h}{T}.$$

For, on integrating with respect to x we obtain

$$\overline{\int_A^B \dot{S}\, dx} = \left[\frac{q}{T}\right]_A^B + \int_A^B \frac{gq}{T^2}\, dx + \int_A^B \frac{h}{T}\, dx,$$

and the Clausius–Duhem inequality follows on invoking the heat conduction inequality.

46. At this point we shall need to make the same definitions as we did in §19. Thus, the rate of working by the body is

$$W = -[\sigma \dot{u}]_A^B,$$

the net heat flux into the body is

$$Q = [q]_A^B,$$

the rate of absorption of heat by the body is

$$Q^+ = \text{Max}(Q, 0),$$

the rate of emission of heat by the body is

$$Q^- = -\text{Min}(Q, 0),$$

and so forth.

If it is the case that

(i) $h = 0$ on $[A, B] \times [t_1, t_2]$,
(ii) $T(A, t) = T(B, t) \ (= \tau(t)$ say) on $[t_1, t_2]$,

then the Clausius–Planck inequality

$$\frac{Q}{\tau} \leqq \overline{\int_A^B \dot{S}\, dx}$$

is valid and if, in addition,

(iii) $[\int_A^B S\, dx]_{t_1}^{t_2} = 0$

then the Clausius inequality

$$\int_{t_1}^{t_2} \frac{Q}{\tau} \, dt \leq 0$$

is valid.

Once again the proof is straightforward. For, when $h = 0$, the argument of §45 shows that

$$\overline{\int_A^B \dot{S} \, dx} = \left[\frac{q}{T}\right]_A^B + \int_A^B \frac{gq}{T^2} \, dx \geq \left[\frac{q}{T}\right]_A^B.$$

Since, as (ii) implies,

$$\left[\frac{q}{T}\right]_A^B = \frac{1}{\tau}[q]_A^B = \frac{Q}{\tau}$$

the Clausius–Planck inequality is correct.

Furthermore, an integration with respect to t yields

$$\int_{t_1}^{t_2} \frac{Q}{\tau} \, dt \leq \left[\int_A^B S \, dx\right]_{t_1}^{t_2}$$

and, thus, the additional assumption (iii) delivers the Clausius inequality.

47. The hypothesis $f = 0$ was not needed in §46 but it is needed now.

(Clausius–Kelvin) *Suppose that*

(i) $f = h = 0$ *on* $[A, B] \times [t_1, t_2]$,
(ii) $T(A, t) = T(B, t) \, (= \tau(t) \; say)$ *on* $[t_1, t_2]$,
(iii) $[\frac{1}{2}\int_A^B \rho \dot{u}^2 \, dx + \int_A^B U \, dx]_{t_1}^{t_2} = 0$,
    $[\int_A^B S \, dx]_{t_1}^{t_2} = 0$,

and suppose, in addition, that there is a positive constant τ_0 such that

(iv) $\tau(t) < \tau_0$ *at every t at which $Q(t) > 0$,*
 (v) $\tau(t) > \tau_0$ *at every t at which $Q(t) < 0$.*

Then

$$\int_{t_1}^{t_2} W \, dt \leq 0.$$

In brief, it is impossible for the work done by the body to be positive if all heat is absorbed at temperatures lower than any temperature at which heat is emitted.

Hypotheses (i), (ii), and (iii) agree with the hypotheses of §20, and so it must be that the work done by the body coincides with the net heat gained by it, that is to say

$$\int_{t_1}^{t_2} W \, dt = \int_{t_1}^{t_2} Q \, dt.$$

Furthermore, the hypotheses are more restrictive than those of §46 and so the Clausius inequality

$$\int_{t_1}^{t_2} \frac{Q}{\tau} \, dt \leqq 0$$

is valid. On the other hand, hypotheses (iv) and (v) imply that

$$\left(\frac{1}{\tau(t)} - \frac{1}{\tau_0} \right) Q(t) \geqq 0$$

and, therefore,

$$0 \geqq \int_{t_1}^{t_2} \frac{Q}{\tau} \, dt \geqq \frac{1}{\tau_0} \int_{t_1}^{t_2} Q \, dt = \frac{1}{\tau_0} \int_{t_1}^{t_2} W \, dt.$$

Since τ_0 is positive, the desired conclusion now follows.

48. Planck's version of a second law rules out the possibility that all the heat absorbed in a cycle can be converted into work done by the body.

(Planck) *Suppose that*

(i) $f = h = 0$ on $[A, B] \times [t_1, t_2]$,
(ii) $T(A, t) = T(B, t) \, (= \tau(t)$ say$)$ on $[t_1, t_2]$,
(iii) $[\frac{1}{2} \int_A^B \rho \dot{u}^2 \, dx + \int_A^B U \, dx]_{t_1}^{t_2} = 0,$
    $[\int_A^B S \, dx]_{t_1}^{t_2} = 0,$

and suppose, in addition, that

$$\int_{t_1}^{t_2} Q^+ \, dt > 0.$$

Then

$$\int_{t_1}^{t_2} W \, dt < \int_{t_1}^{t_2} Q^+ \, dt, \qquad \int_{t_1}^{t_2} Q^- \, dt > 0.$$

These conclusions follow from the efficiency estimate of §20, that is from the inequality

$$\int_{t_1}^{t_2} W \, dt \leq \left(1 - \frac{m}{M}\right) \int_{t_1}^{t_2} Q^+ \, dt,$$

where the positive numbers M and m are the maximum and minimum values attained by $\tau(t)$ on the interval $[t_1, t_2]$. Since

$$1 - \frac{m}{M} < 1,$$

it must be that

$$\int_{t_1}^{t_2} W \, dt < \int_{t_1}^{t_2} Q^+ \, dt$$

if

$$\int_{t_1}^{t_2} Q^+ \, dt > 0,$$

and since

$$\int_{t_1}^{t_2} W \, dt = \int_{t_1}^{t_2} Q \, dt = \int_{t_1}^{t_2} Q^+ \, dt - \int_{t_1}^{t_2} Q^- \, dt$$

it must also be the case that

$$\int_{t_1}^{t_2} Q^- \, dt > 0.$$

Part 2. Homogeneous and Dissipationless Thermoelasticity

49. The theory of Part 1 of Chapter 2 is presumed to be in force throughout this section. Thus, the external rate-of-heating density

$$h = T\dot{S},$$

where h, T, S are all independent of x, and the rate of heating of the body is

$$H = (B - A)h.$$

It follows trivially that:

The equation

$$(B - A)\dot{S} = \frac{H}{T}$$

holds, and if

$$[S]_{t_1}^{t_2} = 0$$

then

$$\int_{t_1}^{t_2} \frac{H}{T}\, dt = 0.$$

The first of these equations may be interpreted as the degenerate common form assumed by both the Clausius–Duhem inequality and the Clausius–Planck inequality in the present context. The final line is the degenerate form taken by the Clausius inequality.

There are counterparts to the Clausius–Kelvin version of a second law (§47) and to the Planck version (§48). These can be proved by means of arguments that are similar to, but easier than, those deployed in §§47 and 48. I leave it to the reader to supply the proofs but would point out that it is necessary to use a result derived in §24, namely the equality

$$\int_{t_1}^{t_2} W\, dt = \int_{t_1}^{t_2} H\, dt$$

of the work done by the body and the net heat gained by it whenever $[U]_{t_1}^{t_2} = 0$.

(Clausius–Kelvin) *Let*

$$[U]_{t_1}^{t_2} = [S]_{t_1}^{t_2} = 0,$$

and suppose, in addition, that there is a positive constant τ_0 such that

(i) *$T(t) < \tau_0$ at every t at which $H(t) > 0$,*
(ii) *$T(t) > \tau_0$ at every t at which $H(t) < 0$.*

Then

$$\int_{t_1}^{t_2} W \, dt \leqq 0.$$

(Planck) *Let*

$$[U]_{t_1}^{t_2} = [S]_{t_1}^{t_2} = 0$$

and

$$\int_{t_1}^{t_2} H^+ \, dt > 0.$$

Then

$$\int_{t_1}^{t_2} W \, dt < \int_{t_1}^{t_2} H^+ \, dt, \qquad \int_{t_1}^{t_2} H^- \, dt > 0.$$

Part 3. Linearized Thermoelasticity

50. In §§50–53 the theory of Part 2 of Chapter 2 is in force. Since, in particular, the approximate second reduced energy equation, namely

$$q' + h = T_0 \dot{S},$$

holds there are no real counterparts to the Clausius–Duhem inequality (§45) or the Clausius inequality (§46) other than the following, in which

$$Q = [q]_A^B, \qquad H = \int_A^B h \, dx.$$

The equation

$$\overline{\int_A^B S \, dx} = \frac{Q + H}{T_0}$$

holds, and if

$$\left[\int_A^B S \, dx \right]_{t_1}^{t_2} = 0$$

then

$$\int_{t_1}^{t_2} (Q + H)\, dt = 0.$$

51. None the less, the Clausius–Kelvin and the Planck versions of a second law remain valid—the latter with a slight additional qualification. The proofs, though, differ in some essential respects from the proofs of §§47 and 48.

The statement of the Clausius–Kelvin version is identical to that of §47.

(Clausius–Kelvin) *Suppose that*

(i) $f = h = 0$ *on* $[A, B] \times [t_1, t_2]$,
(ii) $T(A, t) = T(B, t)$ $(= \tau(t)$ *say*$)$ *on* $[t_1, t_2]$,
(iii) $[\frac{1}{2}\int_A^B \rho \dot{u}^2\, dx + \int_A^B U\, dx]_{t_1}^{t_2} = 0$,
     $[\int_A^B S\, dx]_{t_1}^{t_2} = 0$,

and suppose, in addition, that there is a positive constant τ_0 such that

(vi) $\tau(t) < \tau_0$ *at every t at which $Q(t) > 0$,*
 (v) $\tau(t) > \tau_0$ *at every t at which $Q(t) < 0$.*

Then

$$\int_{t_1}^{t_2} W\, dt \leqq 0.$$

Hypotheses (i), (ii), and (iii) agree with the hypotheses of §32. Thus the results of that section tell us that

$$\int_{t_1}^{t_2} Q\, dt = 0,$$

$$\int_{t_1}^{t_2} W\, dt + \frac{1}{T_0} \int_{t_1}^{t_2} \int_A^B gq\, dx\, dt = \frac{1}{T_0} \int_{t_1}^{t_2} \tau Q\, dt,$$

and, by virtue of the heat conduction inequality, that

$$\int_{t_1}^{t_2} W\, dt \leq \frac{1}{T_0} \int_{t_1}^{t_2} \tau Q\, dt.$$

However, (iv) and (v) ensure that

$$(\tau(t) - \tau_0)Q(t) \leqq 0$$

and, therefore,

$$\int_{t_1}^{t_2} \tau Q \, dt \leqq \tau_0 \int_{t_1}^{t_2} Q \, dt = 0.$$

Hence

$$\int_{t_1}^{t_2} W \, dt \leqq 0,$$

as required.

52. Within linearized thermoelasticity the Planck version becomes:

(Planck) *Suppose that*

(i) $f = h = 0$ *on* $[A, B] \times [t_1, t_2]$,
(ii) $T(A, t) = T(B, t) \ (= \tau(t) \ say)$ *on* $[t_1, t_2]$,
(iii) $[\frac{1}{2}\int_A^B \rho \dot{u}^2 \, dx + \int_A^B U \, dx]_{t_1}^{t_2} = 0$,
   $[\int_A^B S \, dx]_{t_1}^{t_2} = 0$,

and suppose, in addition, that

$$\int_{t_1}^{t_2} Q^+ \, dt > 0$$

and that M and m, the maximum and minimum values that $\tau(t)$ attains on $[t_1, t_2]$, satisfy

$$M - m < T_0.$$

Then

$$\int_{t_1}^{t_2} W \, dt < \int_{t_1}^{t_2} Q^+ \, dt, \qquad \int_{t_1}^{t_2} Q^- \, dt > 0.$$

It is, of course, implicit in the derivation of the linearized theory from the nonlinear theory that $\tau(t)$ stays close to the reference temperature T_0. Thus, the inequality $M - m < T_0$ entails no real restriction beyond those that have already been made tacitly.

Once again, the proof is straightforward. For, as §32 tells us, the efficiency estimate

$$\int_{t_1}^{t_2} W \, dt \leqq \frac{(M - m)}{T_0} \int_{t_1}^{t_2} Q^+ \, dt$$

holds and, therefore,

$$\int_{t_1}^{t_2} W \, dt < \int_{t_1}^{t_2} Q^+ \, dt$$

if $M - m < T_0$ and

$$\int_{t_1}^{t_2} Q^+ \, dt > 0.$$

Section 32 also tells us that

$$\int_{t_1}^{t_2} Q \, dt = 0$$

and, therefore,

$$\int_{t_1}^{t_2} Q^+ \, dt = \int_{t_1}^{t_2} Q^- \, dt.$$

Thus, it is certainly the case that the right-hand side is positive if the left-hand side is so.

53. I turn now to a different type of formulation of a second law that is due to Perrin; unlike the Clausius–Kelvin version and the Planck version it makes no direct reference to the work done by the body or to the heat absorbed by it.

(Perrin)   *Let the stress–temperature modulus* $\mu(x)$, *the specific heat* $c(x)$, *and (as always) the thermal conductivity* $k(x)$ *all be positive on* $[A, B]$. *Let the body be isolated from its exterior in the sense that*

(i) $f = h = 0$ *on* $[A, B] \times [t_1, t_2]$,
(ii) $\dot{u}(A, t) = \dot{u}(B, t) = q(A, t) = q(B, t) = 0$ *on* $[t_1, t_2]$,

and let

(iii) $\left[\frac{1}{2} \int_A^B \rho \dot{u}^2 \, dx + \int_A^B U \, dx\right]_{t_1}^{t_2} = 0$.

Then the displacement field must be static, and the temperature field must be static and spatially homogeneous, that is to say

$$\dot{u} = \dot{T} = T' = 0 \qquad on \quad [A, B] \times [t_1, t_2].$$

Here isolation of the body is taken to mean that the external force density and the external rate-of-heating density vanish, while the boundary displacements are static and the boundary is thermally insulated.

Hypothesis (iii) would certainly be satisfied if

$$u(x, t_1) = u(x, t_2),$$

$$\dot{u}(x, t_1) = \dot{u}(x, t_2),$$

$$T(x, t_1) = T(x, t_2)$$

at every point x of $[A, B]$. These conditions might be interpreted as saying that the states of the body at the times t_1 and t_2 coincide but, of course, (iii) is a much weaker restriction than is coincidence of the initial and final states.

In order to arrive at the conclusion, we begin by arguing on the same lines as in §33. Thus, Gibbs's relation

$$\dot{U} = \sigma \dot{E} + T \dot{S}$$

is valid (§17) and on adding the term $\rho \dot{u} \ddot{u}$ to each side, and taking account of the momentum equation, which, because $f = 0$, is

$$\sigma' = \rho \ddot{u},$$

we obtain the identity

$$\rho \dot{u} \ddot{u} + \dot{U} = (\sigma \dot{u})' + T \dot{S}.$$

Next, we use the approximate second reduced energy equation, which, because $h = 0$, is

$$q' = T_0 \dot{S}$$

and permits us to substitute q'/T_0 for $\dot{S}$. Thus, we obtain the equation

$$\rho \dot{u} \ddot{u} + \dot{U} = (\sigma \dot{u})' + \frac{T}{T_0} q' = (\sigma \dot{u})' + \frac{1}{T_0}(Tq)' - \frac{gq}{T_0}$$

and, on integrating with respect to x and appealing to the boundary conditions (ii), we find

$$\frac{1}{2} \int_A^B \rho \dot{u}^2 \, dx + \int_A^B U \, dx = -\frac{1}{T_0} \int_A^B gq \, dx.$$

A further integration with respect to t yields the conclusion

$$\int_{t_1}^{t_2} \int_A^B gq \, dx \, dt = 0$$

if (iii) is satisfied. Thus, in view of the heat conduction inequality, it must be that the temperature gradient $g = 0$, that is to say

$$T' = 0 \quad \text{on} \quad [A, B] \times [t_1, t_2].$$

Hence T can depend upon t at most and we may write it as $T(t)$.

The second of the linearized displacement–temperature equations, namely

$$[kT']' = T_0 \mu \dot{u}' + c\dot{T},$$

and the hypothesis that μ is positive, now enable us to conclude that

$$\dot{u}'(x, t) = -\frac{c(x)\dot{T}(t)}{T_0 \mu(x)}.$$

Since $\dot{u}(A, t) = \dot{u}(B, t) = 0$ it must be that

$$\frac{-\dot{T}(t)}{T_0} \int_A^B \frac{c(x)}{\mu(x)} dx = \int_A^B \dot{u}'(x, t)\, dx = [\dot{u}]_A^B = 0.$$

However, the integral

$$\int_A^B \frac{c}{\mu}\, dx$$

is positive if c and μ are positive and, accordingly, $\dot{T}(t) = 0$ and $\dot{u}'(x, t) = 0$. Lastly, the boundary condition $\dot{u}(A, t) = 0$ implies that

$$\dot{u}(x, t) = \int_A^x \dot{u}'(y, t)\, dy = 0$$

and so the proof is complete.

It should be noted that some restriction upon the stress–temperature modulus μ is essential if the result is to be valid. For, if $\mu = 0$ the linearized displacement–temperature equations degenerate to

$$[\beta(x)u']' = \rho(x)\ddot{u},$$

$$[k(x)T']' = c(x)\dot{T}.$$

If $v(x)$ is a characteristic function, and $\omega/2\pi$ is a characteristic frequency, of the Sturm–Liouville problem

$$[\beta(x)v']' = -\omega^2 \rho(x)v, \quad v(A) = v(B) = 0,$$

and if

$$u(x, t) = v(x) \cos \omega t, \quad T(x, t) = T_0,$$

then u and T satisfy the linearized displacement–temperature equations, and $\dot{u}$ and q vanish at the boundary points. Moreover,

$$\left[\tfrac{1}{2} \int_A^B \rho \dot{u}^2 \, dx + \int_A^B U \, dx \right]_{t_1}^{t_2} = 0$$

if $t_2 - t_1 = 2\pi/\omega$, but u is not static even though T is static and spatially homogeneous.

Part 4. Nonstandard Linearized Thermoelasticity

54. The preceding section shows that Perrin's version of a second law is indeed a logical consequence of the equations of linearized thermoelasticity. It would seem that Perrin believed that his version exhausts all the content of a second law but it is the purpose of this section and the next to point out that this cannot be the case.

I propose to work within the context of nonstandard linearized thermoelasticity, which is slightly more general than the linearized theory set out in Part 2 of Chapter 2.

In order to arrive at this theory, let us return to the momentum equation

$$\sigma' + f = \rho \ddot{u},$$

and the first reduced energy equation

$$\sigma \dot{E} + q' + h = \dot{U}.$$

These equations depend only upon the momentum balance law and the energy balance law. The derivation of the linearized theory was made to depend upon the second reduced energy equation, which requires the equations of state to be satisfied, but the equations of state are not now presumed to be in force.

We now impose the simplification:

There is a positive constant T_0 such that $\hat{\sigma}(0, T_0, x)$ is independent of x.

The same simplification was made in §13; as before, T_0 is the reference temperature and the constant

$$\sigma_0 = \hat{\sigma}(0, T_0, x)$$

is the residual stress.

We depart from §14 in that rather than replace the term $T\dot{S}$ by the term $T_0\dot{S}$ in the second reduced energy equation we adopt the approximation:

In the first reduced energy equation, the term

$$\sigma\dot{E}$$

may be replaced by

$$\sigma_0\dot{E},$$

and, therefore, the approximate first reduced energy equation

$$\sigma_0\dot{E} + q' + h = \dot{U}$$

is presumed to be in force.

If, as in §15, it is supposed that E, $T - T_0$, g are small, it is natural to attempt to approximate the response functions for the stress, the internal energy density, and the heat flux, by way of the relations

$$\hat{\sigma}(E, T, x) = \sigma_0 + \partial_E\hat{\sigma}(0, T_0, x)E + \partial_T\hat{\sigma}(0, T_0, x)(T - T_0),$$

$$\hat{U}(E, T, x) = \hat{U}(0, T_0, x) + \partial_E\hat{U}(0, T_0, x)E + \partial_T\hat{U}(0, T_0, x)(T - T_0),$$

$$\hat{q}(E, T, g, x) = \hat{k}(0, T_0, x)g.$$

The following notations and names are employed for the coefficients:

   the residual internal energy density $U_0(x) = \hat{U}(0, T_0, x)$,
   the isothermal elastic modulus $\beta(x) = \partial_E\hat{\sigma}(0, T_0, x)$,
   the stress–temperature modulus $\mu(x) = -\partial_T\hat{\sigma}(0, T_0, x)$,
   the specific heat at constant strain $c(x) = \partial_T\hat{U}(0, T_0, x)$,
   the thermal conductivity $k(x) = \hat{k}(0, T_0, x)$.

Within linearized thermoelasticity the coefficient

$$\partial_E\hat{U}(0, T_0, x) = \sigma_0 + \mu(x)T_0,$$

as is clear from §17, but we do not now adopt this equation. Instead we introduce an unnamed modulus

$$v(x) = (\partial_E\hat{U}(0, T_0, x) - \sigma_0)/T_0,$$

and in this way we arrive at the second, and final, approximation of the nonstandard linearized theory:

The constitutive relations for the stress, the internal energy density, and the heat flux, may be replaced by

$$\hat{\sigma}(E, T, x) = \sigma_0 + \beta(x)E - \mu(x)(T - T_0),$$

$$\hat{U}(E, T, x) = U_0(x) + (\sigma_0 + T_0 v(x))E + c(x)(T - T_0),$$

$$\hat{q}(E, T, g, x) = k(x)g.$$

As in §15, it is the case that:

The heat conduction inequality

$$gq > 0 \quad \text{if} \quad g \neq 0$$

is satisfied if and only if

$$k(x) > 0 \quad \text{in } [A, B],$$

and this last restriction is presumed to be in force.

If we substitute from the approximate constitutive relations into the momentum equation and the approximate first reduced energy equation, and remember that $E = u'$ and $g = T'$, we arrive at the conclusion:

Within nonstandard linearized thermoelasticity the displacement and temperature fields satisfy the nonstandard linearized displacement–temperature equations

$$[\beta u' - \mu(T - T_0)]' + f = \rho \ddot{u},$$

$$[kT']' + h = T_0 v u' + c\dot{T}.$$

The first nonstandard equation is identical to its counterpart in the linearized theory, but the second differs from its counterpart in that the term

$$T_0 v \dot{u}'$$

replaces the term

$$T_0 \mu \dot{u}'.$$

Within the linearized theory the identity

$$\mu(x) = v(x) \quad \text{on } [A, B]$$

holds; that it does so is a consequence of Maxwell's relation (§9).

55. It is proposed to operate by not presupposing the coincidence of μ and v and asking whether Perrin's version of a second law implies their coincidence. It will be found that this is not so for certain affine relations between μ and v are consistent with Perrin's version.

Let the moduli $\mu(x)$, $v(x)$, the specific heat $c(x)$, and (as always) the thermal conductivity $k(x)$, all be positive on $[A, B]$ and suppose that there are constants a and b such that

$$\mu(x) = a + bv(x),$$

where a is sufficiently small as to ensure that

$$a^2 \int_A^B \frac{v}{k(a + bv)}\, dx \cdot \int_A^B \frac{kv'^2}{v^3(a + bv)}\, dx < 1.$$

Let the body be isolated from its exterior in the sense that

(i) $f = h = 0$ on $[A, B] \times [t_1, t_2]$,
(ii) $\dot{u}(A, t) = \dot{u}(B, t) = q(A, t) = q(B, t) = 0$ on $[t_1, t_2]$,

and let

(iii) $[\frac{1}{2}\int_A^B \rho\dot{u}^2\, dx + \int_A^B V\, dx]_{t_1}^{t_2} = 0$,
$[\chi]_{t_1}^{t_2} = 0$,

where

$$V(x, t) = \frac{1}{2}\left(\beta E^2 + \frac{c\mu}{vT_0}(T - T_0)^2 \right),$$

$$\chi(t) = \int_A^B \frac{c}{v}(T - T_0)\, dx \bigg/ \int_A^B \frac{c}{v}\, dx.$$

Then the displacement field must be static, and the temperature field must be static and spatially homogeneous, that is to say

$$\dot{u} = \dot{T} = T' = 0 \qquad on \quad [A, B] \times [t_1, t_2].$$

We remark, as we did in §53, that hypothesis (iii) is satisfied if the initial and final states of the body coincide, that is if

$$u(x, t_1) = u(x, t_2),$$

$$\dot{u}(x, t_1) = \dot{u}(x, t_2),$$

$$T(x, t_1) = T(x, t_2)$$

at every point of $[A, B]$, but these conditions are much more restrictive than is (iii).

To prove the assertion we start by multiplying the first of the nonstandard displacement–temperature equations through by $\dot{u}$, and rearranging the expression that results, to obtain

$$(\sigma \dot{u})' = \rho \dot{u} \ddot{u} + (\beta u' - \mu(T - T_0))\dot{u}'.$$

On multiplying the second of the nonstandard equations by $\mu(T - T_0)/vT_0$ and rearranging we find that

$$\frac{1}{T_0}\left[\frac{\mu}{v}(T - T_0)q\right]' = \frac{k\mu}{T_0 v}T'^2 + \left(\frac{\mu}{v}\right)'\frac{k(T - T_0)T'}{T_0}$$

$$+ (T - T_0)\mu \dot{u}' + \frac{c\mu}{T_0 v}(T - T_0)\dot{T}.$$

If we add these equations the term $\mu(T - T_0)\dot{u}'$ cancels, and when we multiply through by T_0 we are left with the identity

$$T_0(\rho \dot{u}\ddot{u} + \dot{V}) + \frac{k\mu}{v}T'^2 + k\left(\frac{\mu}{v}\right)'(T - T_0)T'$$

$$= \left(T_0\sigma \dot{u} + \frac{\mu}{v}(T - T_0)q\right)'.$$

Thus, if we integrate with respect to x and appeal to the boundary conditions we see that

$$T_0\left(\frac{1}{2}\int_A^B \rho \dot{u}^2\, dx + \int_A^B V\, dx\right) + \int_A^B \frac{k\mu}{v}T'^2\, dx$$

$$+ \int_A^B k\left(\frac{\mu}{v}\right)'(T - T_0)T'\, dx = 0.$$

The last term on the left-hand side is troublesome. To deal with it we observe that if we divide both sides of the second nonstandard displacement–temperature equation by v and rearrange we obtain

$$\left(\frac{1}{v}q\right)' = \left(\frac{1}{v}\right)'kT' + T_0\dot{u}' + \frac{c}{v}\dot{T}.$$

Integration with respect to x and an appeal to the boundary conditions now yield

$$\int_A^B \left(\frac{1}{v}\right)'kT'\, dx + \int_A^B \frac{c}{v}\dot{T}\, dx = 0$$

and, hence, the definition of $\chi(t)$ implies that

$$\int_A^B \left(\frac{1}{v}\right)' kT' \, dx = -\dot{\chi} \int_A^B \frac{c}{v} \, dx.$$

On the other hand, the affine relation $\mu = a + bv$ ensures that

$$\left(\frac{\mu}{v}\right)' = a\left(\frac{1}{v}\right)'$$

and, hence, that the integral

$$\int_A^B k\left(\frac{\mu}{v}\right)' (T - T_0) T' \, dx$$

$$= a \int_A^B \left(\frac{1}{v}\right)' k(T - T_0) T' \, dx$$

$$= a \int_A^B \left(\frac{1}{v}\right)' k(T - T_0 - \chi) T' \, dx + a\chi \int_A^B \left(\frac{1}{v}\right)' kT' \, dx$$

$$= a \int_A^B \left(\frac{1}{v}\right)' k(T - T_0 - \chi) T' \, dx - a\chi\dot{\chi} \int_A^B \frac{c}{v} \, dx.$$

Thus we have established the equation

$$\overline{T_0\left(\tfrac{1}{2}\int_A^B \rho\dot{u}^2 \, dx + \int_A^B V \, dx\right) - \tfrac{1}{2}a\chi^2 \int_A^B \frac{c}{v} \, dx}$$

$$+ \int_A^B \frac{k\mu}{v} T'^2 \, dx + a \int_A^B \left(\frac{1}{v}\right)' k(T - T_0 - \chi) T' \, dx = 0.$$

The definition of χ as a weighted mean of the temperature difference $T - T_0$ ensures the existence of at least one point $\xi(t)$ in $[A, B]$
such that

$$\chi(t) = T(\xi(t), t) - T_0.$$

Hence

$$T(x, t) - T_0 - \chi(t) = T(x, t) - T(\xi(t), t) = \int_{\xi(t)}^x T'(y, t) \, dy$$

and Schwarz's inequality yields the estimate

$$|T - T_0 - \chi|^2 \le \left(\int_A^B |T'| \, dx\right)^2 \le \int_A^B \frac{v}{k\mu} \, dx \cdot \int_A^B \frac{k\mu}{v} T'^2 \, dx.$$

Thus

$$\left| \int_A^B \left(\frac{1}{v}\right)' k(T - T_0 - \chi)T' \, dx \right|^2$$

$$\leq \int_A^B \frac{v}{k\mu} \, dx \cdot \int_A^B \frac{k\mu}{v} T'^2 \, dx \cdot \left(\int_A^B k \left| \left(\frac{1}{v}\right)' T' \right| dx \right)^2$$

and, by a further application of Schwarz's inequality to the third integral in the product which appears on the right-hand side of the inequality, we have

$$\left| \int_A^B \left(\frac{1}{v}\right)' k(T - T_0 - \chi)T' \, dx \right|$$

$$\leq \int_A^B \frac{v}{k\mu} \, dx \cdot \int_A^B \frac{kv}{\mu} \left| \left(\frac{1}{v}\right)' \right|^2 dx \cdot \left(\int_A^B \frac{k\mu}{v} T'^2 \, dx \right)^2.$$

In this way we have arrived at the inequality

$$T_0 \left(\frac{1}{2} \int_A^B \rho \dot{u}^2 \, dx + \int_A^B V \, dx \right) - \frac{1}{2} a\chi^2 \int_A^B \frac{c}{v} \, dx + \delta \int_A^B \frac{k\mu}{v} T'^2 \, dx \leqq 0,$$

in which the constant

$$\delta = 1 - |a| \left(\int_A^B \frac{v}{k\mu} \, dx \cdot \int_A^B \frac{kv}{\mu} \left| \left(\frac{1}{v}\right)' \right|^2 dx \right)^{1/2}$$

$$= 1 - |a| \left(\int_A^B \frac{v}{k(a + bv)} \, dx \cdot \int_A^B \frac{kv'^2}{v^3(a + bv)} \, dx \right)^{1/2}$$

is strictly positive. On integrating both sides of the inequality with respect to t, and using hypothesis (iii), we deduce that

$$\int_{t_1}^{t_2} \int_A^B \frac{k\mu}{v} T'^2 \, dx \, dt \leqq 0.$$

The proof can now be completed on much the same lines as in §53. For we have

$$T' = 0 \quad \text{on} \quad [A, B] \times [t_1, t_2].$$

Hence T is independent of x and the second nonstandard equation implies that

$$\dot{u}'(x, t) = -\frac{c(x)\dot{T}(t)}{T_0 v(x)}.$$

Since $\dot{u}(A, t) = \dot{u}(B, t) = 0$ it must be that

$$-\frac{\dot{T}(t)}{T_0} \int_A^B \frac{c(x)}{v(x)} \, dx = \int_A^B \dot{u}'(x, t) \, dx = [\dot{u}]_A^B = 0.$$

Thus $\dot{T} = 0$ and $\dot{u}' = 0$. Finally, we have

$$\dot{u}(x, t) = \int_A^x \dot{u}'(y, t) \, dy = 0$$

and so $\dot{u} = 0$ and the proof is complete.

References

[1] CARLSON, D. E. Linear thermoelasticity. In *Handbuch der Physik*, vol. Vla/2, edited by C. Truesdell. Berlin, Springer-Verlag, 1972.

[2] CARNOT, S. *Reflections on the Motive Power of Fire*, edited by E. Mendoza. New York, Dover, 1960.

[3] COLEMAN, B. D. and W. NOLL. The thermodynamics of elastic materials with heat conduction and viscosity. *Arch. Rational Mech. Anal.* **13**, 167–178 (1963).

[4] COLEMAN, B. D. and D. R. OWEN. A mathematical foundation for thermodynamics. *Arch. Rational Mech. Anal.* **54**, 1–104 (1974).

[5] DAY, W. A. *The Thermodynamics of Simple Materials with Fading Memory*. Springer Tracts in Natural Philosophy, Vol. 22. Berlin, Springer-Verlag, 1972.

[6] LORD KELVIN. An account of Carnot's theory of the motive power of heat. In *Mathematical and Physical Papers of Sir William Thomson*, Vol. I. Cambridge, Cambridge University Press, 1882.

[7] LORD KELVIN. On the dynamical theory of heat. In *Mathematical and Physical Papers of Sir William Thomson*, Vol. I. Cambridge, Cambridge University Press, 1882.

[8] MÜLLER, I. *Thermodynamics*. Boston, Pitman, 1985.

[9] OWEN, D. R. *A First Course in the Mathematical Foundations of Thermodynamics*. Berlin, Springer-Verlag, 1984.

[10] PERRIN, J.-B. Le contenu essentiel des principes de la thermodynamique. In *Oeuvres Scientifiques de Jean Perrin*. Paris, CNRS, 1950.

[11] PLANCK, M. *Treatise on Thermodynamics* (fifth edition). London, Longmans, 1927.

[12] SERRIN, J. (editor). *New Perspectives in Thermodynamics*. Berlin, Springer-Verlag, 1986.

[13] TRUESDELL, C. *The Tragicomical History of Thermodynamics 1822–1854*. New York, Springer-Verlag, 1980.

[14]   TRUESDELL, C. *Rational Thermodynamics* (second edition). New York, Springer-Verlag, 1984.
[15]   TRUESDELL, C. and S. BHARATHA. *The Concepts and Logic of Classical Thermodynamics as a Theory of Heat Engines.* New York, Springer-Verlag, 1977.

Index

Printed in the United States
by Baker & Taylor Publisher Services